Robust Simulation for Mega-Risks

Craig E. Taylor

Robust Simulation for Mega-Risks

The Path from Single-Solution to Competitive, Multi-Solution Methods for Mega-Risk Management

Springer

Craig E. Taylor (deceased)
ImageCat, Inc.
Torrance, CA, USA

ISBN 978-3-319-19412-7 ISBN 978-3-319-19413-4 (eBook)
DOI 10.1007/978-3-319-19413-4

Library of Congress Control Number: 2015946985

Springer Cham Heidelberg New York Dordrecht London

Printed on acid-free paper

Springer International Publishing AG Switzerland is part of Springer Science+Business Media (www.springer.com)

Dedicated to Dr. Craig E. Taylor
(1945–2014)
Vires Acquirit Eundo
"We gather strength as we go"
The Aeneid by Virgil

Credits

Principal Author
Craig E. Taylor

Advisory Team
Melissa Taylor Dresler
Yajie Lee
Adam F. Taylor

Contributors
William P. Graf
Charles K. Huyck
Zhenghui Hu
Ronald T. Eguchi
Robert Riehemann
David Perkins
Abigail Horn

Special Thanks to the full team at ImageCat
Yajie Lee
William P. Graf
Charles K. Huyck
Zhenghui Hu
Ronald T. Eguchi

Additional Thanks
Fritz Schmuhl
Adam Rose
Syed Rashid Minhas
Gayle Taylor
Romy Taylor
Catherine Taylor
Slava Petuhov
Madison Dresler

Foreword

Natural and human-made events, such as earthquakes, hurricanes, epidemics, and terrorist attacks, can cause colossal damage and immense social and economic disruptions. These threats are never completely avoidable, but often can be mitigated or preempted if the necessary resources are allocated and prudent actions are taken in advance. Preparations can include retiring or retrofitting existing vulnerable facilities, researching and enhancing the effectiveness of construction codes, planning future developments with conscious risk considerations, and utilizing financial vehicles to transfer losses and fund post-disaster recovery. In reality, however, resources are always limited, forcing decision-makers to make judgments about priorities based on the information at hand. Preparing for potential future events can be seen as less pressing than addressing more substantive current crises if the potential magnitude of these events is not fully understood. Decision-makers need more information to assist in creating plans and appropriately allocating available resources before an event strikes in order to build a more robust and disaster-resilient community.

Achieving this goal begins with a realistic assessment of a community's potential disaster-related risks. This, however, is by no means an easy task. Assessing risks includes understanding and modeling the very complex 1) physical process of the potential risk events, 2) response of the community as a whole as well as its individual systems, and 3) potential socioeconomic consequences. In the past several decades, numerous research studies have been performed across a broad range of disciplines to assess potential impacts from mega-risk events. Scientific models and analytical frameworks are continually developed. Methodologies undergo ongoing improvement with advances in subdisciplines. These methodologies, unfortunately, are generally mathematically driven and computationally intensive. The wide use of computer technology from early 1980s has introduced a revolutionary change in how mega-risks can be studied and analyzed. Today, with continuous and rapid advancement in computing technologies, risk modeling techniques are improving at an unprecedented speed.

That said, challenges are ubiquitous. Fundamental questions on the modeling techniques remain. Mega-risk events often involve small probability but large consequences. Documentation of past mega-events is very young compared to the cyclical timelines behind these mega-risk events. Historical evidences are often inferred and subject to substantial uncertainties. Scientific models, both empirical and physics-based, are developed often without sufficient understanding of the physical process and the associated socioeconomic consequences of events. Statistics are used heavily. Models may be constructed based on different, partial or full data sets that are available to different researchers (or research groups), or sometimes even the same researcher (or group). These models may all fit the data set used for development to various degrees with acceptable criterion (or criteria). Theories and assumptions behind the models, however, can be quite different, or even contradictory. As a result, divergent outcomes may be produced when modeling risks from future events. Furthermore, traditional statistical and probability theories and methods, many developed with experiences from large numbers of samples and assuming "stable" distribution, are often "blindly" applied to solve mega-risk problems, which by nature are small probability and must be modeled with extreme distributions. These extreme distributions are typically much less stable and more difficult to calibrate. Furthermore, the large uncertainty intrinsically associated with the highly sophisticated analytics is often hidden, either voluntarily or intentionally, with most contemporary modeling approaches. Single solutions are commonly provided to decision-makers, with illusory precision.

A pioneer and highly respected scholar of mega-risk management, Dr. Craig Taylor, has included a series of essays in his book, *Robust Simulation for Mega-Risks*, which provide an in-depth overview of the history and perspectives of the traditional theories and methods for statistics and probability. These schools of thought are the foundation of modern risk assessment. In his essays, Dr. Taylor explores their individual weaknesses and limitations, and in particular the danger of "blindly" extrapolating to solve mega-risk problems. By looking into the examples of many of the significant findings that played major roles in human history and their evolution from the initial discovery to a final resolution to major disasters, Dr. Taylor explains why divergent answers from competitive approaches to mega-risk problems are not only a necessity but are also unavoidable. A simplistic, forced consensus or single-answer view can not only be misleading and seriously skew the "true" picture of mega-risks, but may also hinder the progress of science and technological advancement. In that regard, outliers in predicted future losses are not only of intellectual interest, but also of great importance to define the "bound" of future risks in a meaningful way. Furthermore, Dr. Taylor lays the foundation for a new analytical approach, called "Robust Simulation," that provides the framework for constructing a more robust view of mega-risks from alternative models. Also discussed are the various analytical methods that can utilize divergent solutions to facilitate more robust financial and risk-mitigation decisions. In his essays, Dr. Taylor defines a new simple metric, called the "Cat Index," to measure the stability of simulated losses or risk exposures in the context of extreme distributions and to answer important questions such as: does the implied loss or exposure distribution have a stable standard deviation, or does it even have a stable mean? All these are of great importance in risk-related applications such as insurance, monetary investment, and more.

With extensive experience and profound thinking, Dr. Taylor sheds light on many innovative ideas and pragmatic approaches for tackling natural and human-made mega-risk issues, which are invaluable to risk management practitioners. In his words, "The industry needs a rethink and paradigm change" on our current single-solution approaches to achieve a more robust view of the actual risks to build more disaster-resistant and safer communities.

ImageCat, Inc.
Long Beach, CA
June 2015

Yajie Lee

Preface

My first email about this book came in April 2013. My father was writing a book and needed my help—someone to read the draft chapters and give feedback and guidance. This first message discussed American psychologist Abraham Maslow, Plato, the fusion of East-West philosophies, and Aristotle's concept of "eudemonia" or happiness in the fullest sense. If you didn't know my father or his style, you might find it unusual that a book about risk and natural hazards would start with philosophy, psychology, religion, and history. But if you knew him you would find it quite normal. He was extremely intellectually curious, believing that knowledge of all topics comes into play in solving complex systems problems.

The starting point for this book was a paper written by my father and several of the colleagues who contributed to this book: Yajie Lee, Bill Graf, Zhenghui Hu, and Charles Huyck all of ImageCat. The paper was written for a 2010 conference in Shanghai and entitled: "Robust Simulation and Cat Diagnostics for Treating Uncertainties in Catastrophe Risk Analysis." In this paper, robust simulation is discussed as a solution for developing a statistical analysis in an area with considerable competition among analysts but where policy or financial decisions are needed in the near term. The paper defines robust simulation as a method that involves selecting a preferred model to run simulations and then selecting alternate models to provide other answers. The range of these competitive answers then would provide the range of uncertainties … "as best we can know them at a given time."[1]

My father had a dominant core idea throughout his career as a seismic risk expert about the importance of looking at multiple outcomes or "ensembles." Modeling a single outcome was insufficient in a world where surprise is common, where outliers could be catastrophic, where the mean is unstable and where the world itself is

[1] Taylor, Craig, Yajie Lee, William Graf, Zhenghui Hu, and Charles Huyck. (2010). "Robust Simulation and Cat Diagnostics for Treating Uncertainties in Catastrophe Risk Analysis," *Reliability Engineering and Risk Management: Proceedings of the International Symposium on Reliability Engineering and Risk Management*, Tongji University Press.

changing. He was keenly aware that this challenged the long-standing tradition in Western thought of looking for single solutions and single outcomes.[2] In the context of seismic risk, this meant modeling not just one likely outcome but a suite of potential outcomes—a stochastic method rather than a deterministic one. The concept of "robust" or "robustness" takes this another step allowing for the consideration of competing strategies. The core idea in this book is that the ensemble view instead of the mean or variance is better capable of characterizing mega-risks. This figure of loss estimates for seismic risk assessment demonstrates how many policy and financial decisions about risk are currently made based on a single or central scenario rather than an ensemble view.[3] Other simulations demonstrate that robust simulation better captures the variability of real world events.[4]

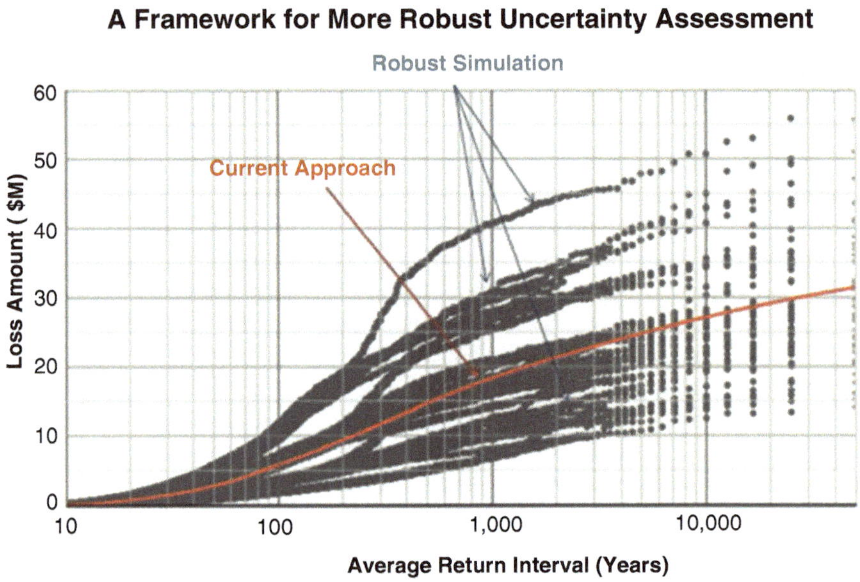

Source: ImageCat (2014)

---

[2] Toulmin, Stephen, (1992). *Cosmopolis: The Hidden Agenda of Modernity*, Chicago: University of Chicago Press.

[3] "Using Robust Simulation to Characterize Uncertainties in Catastrophe Loss Assessments," from RAA Cat Modeling 2014, ImageCat, Inc.

[4] See Taylor, Craig, William Graf, Yajie Lee, Charles Huyck, and Zhengui Hu, 2012, "Propagation of Uncertainties through Robust Simulation and Future Research," *Fifth Asian-Pacific Symposium on Structural Reliability and its Applications (5APSSRA)*. Phook, K. K., Beer, M., Quek, S. T. and Pang, S. D., editors, Singapore.

In this book, robust simulation is introduced in Chap. 7. In fact, my father had many working titles and subtitles for the book itself which did not even contain either of those words—instead referencing multiple interpretations, paths, risks, and uncertainties. He liked the notion of mega-risks, a way to expand his ideas about decision strategies to important topics beyond earthquakes. My father also wrote the chapters in sequence, first writing about deductivist theory, then frequentist theory, and then subjectivist theory—each ultimately judged to be valuable but insufficient for the problem at hand. As he sent me each chapter, I would ask him if he would ever propose something affirmative—something new perhaps? I told him that the book should propose a new methodology—people like that in a book. But in fact, this was the plan all along if only I had the patience. So the book is about process, the path to robust simulation, walking through the key schools of thought and conceptual quandaries that brought us here. In this quest he wanted to be reverential of past and alternate approaches, as he quotes the Hungarian philosopher Imre Lakatos: "Important criticism is always constructive: there is no refutation without a better theory."[5] To this end, the original preface is now the essay in the back on "Learning from Tradition"—a wonderful piece about the trajectory of the book but overtaken by events as a preface.

My father did not live to see the completion of this book. My brother Adam and I were blessed to see my father May 10, 2014 for a lengthy session where the three of us collectively figured out the flow of the book, how the chapters came together as a manuscript. My father wanted us to finish his book without him and we have attempted to keep it as close as possible to his own wording and voice. My father was a rigorous, critical thinker. Although he is not here to see this book published, he would want readers to approach this book with critical and inquisitive minds. We do not present this out of sentimentality but out of the importance and uniqueness of the ideas and as a significant contribution to the field. My father would welcome all competitive, divergent, alternate, nonlinear views and interpretations. It's a book about eudemonia, here a desire to address important risks facing the world, and an understanding that this approach would represent one moment in time until it helps propel new thinkers with new questions and new ideas.

Many people have helped with this book. Three of us have served as an editorial and writing team after we lost my father: myself, Adam Taylor, and Yajie Lee who has helped joyfully with every question that has arisen. Enormous thanks goes to the full team at ImageCat: in addition to Yajie, we thank William Graf, Charles Huyck, Zhenghui Hu, and Ron Eguchi. Adam Rose was a cherished friend and colleague and helped with review and publication. Robert Riehemann has provided enormous assistance and extremely challenging comments. Abigail Horn has been a tremendous help and brings amazing resources to our project. Syed Rashid Minhas

[5] From p. 6, Lakatos, I., 1978, *The methodology of scientific research programmes*, London: Cambridge University Press.

provided outstanding technical support. We thank the team at Springer—Fritz Schmuhl, who has been so patient and helpful, Timo Bazuin, and Naomi Portnoy. Thank you Romy and Cassie. Madison Dresler has provided excellent editorial support. Thank you Gayle Taylor for your grace, kindness, courage, and strength and for understanding the importance of fulfilling this promise.

Natural Hazards Management, Inc. Melissa Taylor Dresler
Torrance, CA
May 2015

Contents

Part I
Introduction and Scope

Chapter 1
Introduction: An Inquiry-Based Approach

NATURE and Nature's laws lay hid in night;
God said, Let Newton be! *and all was light. (From "EPITAPH, Intended for Sir Isaac Newton," by Alexander Pope, 1730, p. 9 in* The Poetry of Pope: A Selection, *edited by M. H. Abrams, New York, Appleton-Century Crofts, Inc., 1954)*

The first and most important ability you can develop in a flat world is the ability to "learn how to learn"—to constantly absorb, and teach yourself, new ways of doing old things or new ways of doing new things. ... Because what you know today will be out-of-date sooner than what you think. (From p. 302, Friedman, Thomas L., 2006, The World is Flat*)*

Abstract This book introduces a new way of analyzing, measuring, and thinking about mega-risks, a "paradigm shift" that moves from single solutions to multiple competitive solutions and strategies. "Robust simulation" is a statistical approach that yields ranges of answers and requires a process or strategy that takes into account alternate competitive evaluations. To reach this approach, the book systematically walks through the historical statistical methods for evaluating risks. The first chapters deal with three theories of probability and statistics that have been dominant in the twentieth century: deductivist, frequentist, and subjectivist. The book then introduces "robust simulation" which solves the problem of measuring the stability of simulated losses, incorporates outliers, and simulates future risk through simulation of a suite of possible answers. The book emphasizes the importance of flexibility and attempts to demonstrate that alternative credible approaches are helpful and required in understanding a great many phenomena. In its approach, the book is neither *assertoric* (asserting that a clear answer exists) nor is it via *negativa* (discrediting alternate theories)—rather it is *inquiry based*, looking systematically at extant theories to collect and distill all relevant lessons.

C.E. Taylor, *Robust Simulation for Mega-Risks*, DOI 10.1007/978-3-319-19413-4_1

1.1 Initial Queries and the Widening Approach

This book begins with the basic question: "How do we account for uncertainties in mega-risk evaluations?" This is different from asking how risk assessment and risk and decision procedures contribute to understanding and coping with mega-risks.[1] Alternate approaches will not be dismissed out of hand but are systematically reviewed to assess their relative contributions to the problem at hand.

This work will focus primarily on those uncertainties for which we can currently provide some credible risk estimates.[2] Not every "unknown" is to be included as an uncertainty in mega-risk estimates. Unknown consequences, those beyond what is currently known, present opportunities and limits on what can be covered in response to this basic initial question.

This work deals with the mega-risks associated with such natural perils as earthquakes and hurricanes, floods, winter storms, wildfires, tsunamis, avalanches, and landslides. Mega-risks, though, can also be connected to a variety of other perils such as epidemics, highly unpredictable costs of mega-construction projects, asteroids, missiles, climate change, civil wars, wars between nations, terrorist acts, conflagrations, oil spills, crop failure, food contamination, and economic panics and disasters. Mega-risks are here characterized as risks of shocks to complex systems the impacts of whose possible occurrence are of great concern to the human community.[3]

The complex systems in question include energy, water, wastewater, transportation, and communication infrastructure; it could also include healthcare delivery, governmental services, legal, financial, manufacturing, labor, management, and regulatory systems, food and dietary services, and systems of individual people and animals.

Mega-risk evaluations typically cover very important social issues—prospective climate change impacts to specific regions; cost effectiveness of safe rooms in Oklahoma City and other high hazard tornado regions; wind and earthquake design levels for new building construction; how well electric power systems can be expected to respond to flooding, winter storms, earthquakes, hurricanes, and other natural hazards; and to what extent crop failures in the Midwest impact the regional economy.

---

[1] This overarching question is an extension of what was proposed by Melissa Taylor Dresler in her review of the fourth draft of these essays.

[2] On pp. 220–225, Bernstein, Peter L., 1996, *Against the Gods: The Remarkable Story of Risk*, New York: John Wiley & Sons, Inc., one finds a discussion of sharp distinctions between risk and uncertainty but also a discussion of what is [currently] definable and what is not. Some risk procedures such as those for construction risk management can be in extremely early stages of development, and probability estimates may be very crude. Other factors in risk procedures may be virtually unknown at present but may only be revealed as knowledge develops considerably.

[3] Many use a quantitative definition of "shock" to mean data two or more standard deviations away from the mean.

1.2 A Brief Account of Outputs Essential in Mega-Risk Evaluations

1.2.1 *Mega-Risk Evaluation Performed Beforehand*

Figure 1.1 helps to illustrate desired outcomes of mega-risk evaluations. Figure 1.1 provides a loss distribution that assists in illustrating desired outcomes of mega-risk evaluations. This loss distribution begins with losses just below 97 % probability of non-exceedance and proceeds cumulatively from there. In effect, there is only slightly above a 3 % chance that some loss will occur. Such a loss distribution can be defined for the status quo of a complex system, or it can be defined for the system as modified through proposed risk-reduction procedures.

Such risk evaluations can be used or modified to provide such statistical figures of merit as:

- Distributions of risks
- Distributions of risk and benefits (net benefits or losses)
- Probability estimates of failure or success

Distributions are here described as a comparison of either probability (and/or frequency of occurrence) and severity. Probability density functions (pdfs) provide such comparisons relative to various levels of severity. Cumulative probability functions (cdfs) sum these severities from low to high relative to probabilities (and/or frequencies) of exceedance. Resulting metrics may include estimates of:

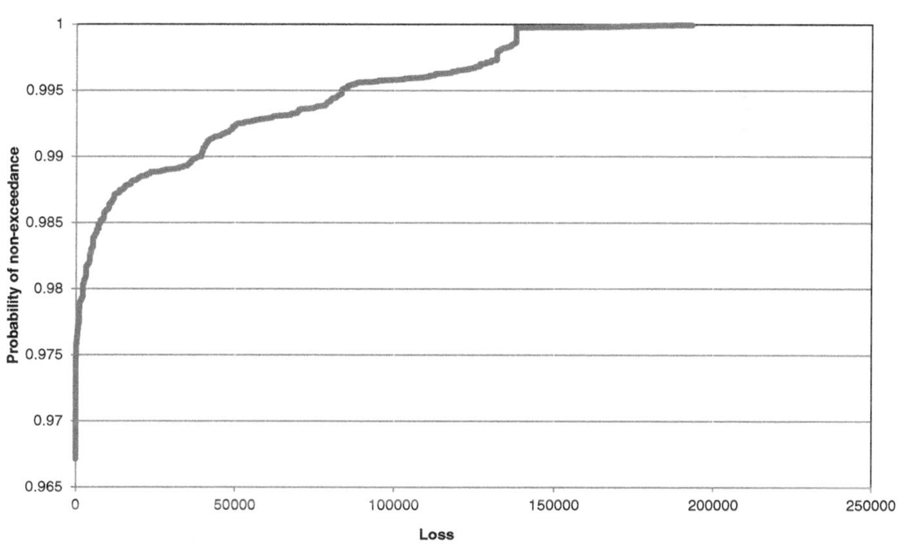

Fig. 1.1 The probability of non-exceedance of a loss level. This figure provides a loss distribution. This loss distribution begins with losses just below 97 % probability of non-exceedance and proceeds cumulatively from there. In effect, there is only slightly above a 3 % chance that some loss will occur

Mean values
Median values
Statistical variances
Severities at different levels of exceedance (e.g., such fractile estimates as the severity at a 1 % annual chance of occurrence or, in frequency terms, with an occurrence on the average every 100 years)

Note that frequencies and probabilities are not equivalent inasmuch as frequencies may exceed one, but probabilities can never exceed one. There may be several accidents in a given year on Hawthorne Blvd., but the probability of an accident in a given year cannot exceed one.

1.2.2 Mega-Risk Evaluations After or During a Major Mega-Risk Event

After a mega-risk event, estimates are typically needed pertaining to losses, injuries, deaths, and also impacts on the environment as well as other complex systems. It is rarely known with certainty what these losses may be. It can be known that a tornado has demolished a house, but estimates of costs to reconstruct this house may vary. Counting injuries may be challenging if many people do not report their injuries. Counting deaths may be nearly impossible if, as in the 2008 Wenchuan, China, earthquake, about 19,000 people who disappeared were suspected of being buried in landslides. Initial estimates of insured losses were $14B after the 1994 Northridge earthquake. Since some claims and their resolution take a very long time to resolve, this estimate has risen to above $19B. On the scientific side, there are not enough sensors to know with certainty the wind velocities and all sites impacted by a hurricane. Inverse modeling of events can also yield various answers.

There are also questions about what to include in the long-term process of a mega-event. For instance, does the Panic of 1907 follow the 1906 San Francisco earthquake, or was this panic partly the result of voluntary investments in rebuilding after the earthquake, investments that could have been far less?[4] Were the Iraqi and Afghanistan wars the consequence of the terrorist acts on the World Trade Center on September 11, 2001, or were these wars voluntary decisions?

Common, then, are ranges of estimates for mega-risk events, and these may improve over time. Not all is necessarily known about mega-events.

---

[4] See Bruner, Robert F and Sean D. Carr, 2007, *The Panic of 1907: Lessons Learning from the Market's Perfect Storm*, Hoboken, N. J.: John Wiley & Sons, Inc.

1.2.3 Uses of These Mega-Risk Estimates

This book will discuss some of the quantitative risk and decision procedures that assist in helping to understand how these statistics can be used. For present purposes, these mega-risk statistics may be conjoined with various decision procedures to assist in specific decisions. For instance, a building code may recommend that, for a new design level, "no collapse" should be expected with a probability of occurrence of greater than 2 % in 50 years. Likewise, one may use benefit-cost evaluations to assign a benefit-cost ratio of 3.1 to the use of sodium hypochlorite to replace chlorine gas in wastewater filtration plants. More advanced financial evaluations can also be employed as needed.

Of special interest are those shocks that give rise to many adverse consequences to systems. These shocks may be rapid like explosions and earthquakes, or very slow as in climate change, or oscillating considerably as in the weather storms, plagues, or chickenpox. Proximate causes may yield large-scale impacts ("tipping points") in some cases, whereas very large events (huge hurricanes) may cause very little damage for extremely wind-resistant cities.[5]

1.3 Book Structure and Organization

1.3.1 Early Chapters: Discussion and Analysis of Seventeenth-to Early Twentieth-Century Approaches to Understanding Risk

Of the many historical advances in probability and statistics for addressing the basic initial question, two are of immediate interest here, and these come from Great Britain in the eighteenth century. First, in his mature works, David Hume[6] maintains that a system of belief, stressing custom and regularity, can explain inductive inferences. We can invoke the uniformity of nature: past instances provide the guide to the similar future. The Newtonian and deterministic view of the world may well be

[5]The expression "tipping point" is popularized in Gladwell, Malcolm, 2002, *The Tipping Point*, New York: Little, Brown and Company. Chaos theory has provided a deterministic version of this view of how small events can give rise to large consequences (see Wikipedia, "Chaos theory," accessed May 20, 2013).

[6]Hume, David, 1748, *An Inquiry Concerning Human Understanding*, Indianapolis: The Bobbs-Merrill Company, Inc., The Library of Liberal Arts, 1955, publication; see Smith, Norman Kemp, 1966, *The Philosophy of David Hume: A Critical Study of Its Origins and Central Doctrines*, New York: St. Martin's Press. In his earlier 1739 work, Hume maintains that "a superior number of chances produces our assent neither by *demonstration* nor by *probability*," p. 126 in Hume, David, 1739, *A Treatise of Human Nature: Being an Attempt to introduce the experimental Method of Reasoning into Moral Subjects*, reprinted Oxford: at the Clarendon Press, 1968. Hume's later theory of belief serves to ameliorate the skepticism from such early remarks.

true and believed in this system. Our experience can, of course, mislead us, as when explorers from Great Britain discovered black swans in Australia and elsewhere.[7] Moreover, in some cases we do not need many instances in order to have a guide as to what to say. Ideas alone don't warrant inductive reasoning. Yet, things work well when custom and regularity are assumed. Joining Hume in a similar ontological climate of events and putative laws, Thomas Bayes developed procedures for augmenting empirical evidence with prior knowledge of some situation. Arguably, Bayes' views provide for more active mental processes to assist inductive inference than are accounted for in Hume's theories.[8]

A discussion of early theories as they have been updated helps to determine how well these theories can evaluate risks to complex systems and accounting for uncertainties in these risk estimates.

Historically we find that some insurance actually covered very extreme risks. For instance, Huebner et al. write that insurance may have begun in Babylonia five or six millennia ago. Caravans subject to thievery and piracy may have been involved.[9] According to Peter Bernstein, the Code of Hammurabi in about 1800 B.C. discussed "bottomry," whereby a ship's owner took out a loan to finance a ship's voyage but did not need to repay the loan if the ship was lost.[10] In Shakespeare's *The Merchant of Venice*, Shylock provides "risk capital" for the merchant Antonio's "argosy bound to Tripolis, another to the Indies." Antonio provides the high-risk "insurance." Speculatively speaking, opulent people such as Portia could then serve as ultimate reinsurers of such ventures.[11] The history of risky distributions was in its very early stages in discussions of insurance and statistics in the sixteenth century and for some period afterward.

Thus, in the eighteenth-century period, we have the important early discussions of what have been still today major approaches to probability and statistics:

The deductivist approach: inductive inferences are merely matters of ideas, of deductive reasoning, and so only require robotic methods for their correct application (covered in Chap. 2).

The frequentist approach: the more evidence that one has, the more confident one can be in one's inferences since asymptotically the evidence yields "laws" or very well-founded inductive inferences based on regularities (covered in Chap. 3).

---

[7] Wikipedia, "Black Swan," accessed May 21, 2013, contains an account of the European discovery of black swans.

[8] From p. 14, Gelman, Andrew, John B. Carlin, Hal S. Stern, and Donald B. Rubin, 2003, *Bayesian Data Analysis*, Boca Raton: Chapman & Hall/CRC. Some original works by Bayes can be found as Appendices in Press, S. J., 1989, *Bayesian Statistics*, New York: John Wiley & Sons.

[9] See, for instance, p. 18 in Huebner, S. S., Kenneth Black, Jr., and Robert S. Cline, 1982, *Property and Liability Insurance*, third edition Englewood Cliffs, N. J.: Prentice-Hall, Inc.

[10] From p. 92 in Bernstein, Peter L., 1996, *Against the Gods: The Remarkable Story of Risk*, New York: John Wiley & Sons, Inc. This work also describes the development of life statistics and other actuarial statistics that, for instance, helped to form Lloyd's of London as early as 1687 (on p. 89).

[11] Shakespeare, William, circa 1595, *The Merchant of Venice*, pp. 578–612 in *Shakespeare: The Complete Works*, edited by G. B Harrison, New York: Harcourt, Brace Jovanovich, Inc.

The subjectivist (Bayesian) approach contends that inductive inferences can be improved if they are augmented by personal or subjective prior estimates (covered in Chap. 4).

In this book, these three approaches arising in the eighteenth century will be updated chiefly by three major figures in the twentieth century: Rudolf Carnap ("the deductivist approach"), Richard von Mises (the "frequentist approach"), and Bruno de Finetti (the "Bayesian approach").[12] These distinguished representatives of the three approaches have each assiduously elaborated their positions and so have dealt with some if not many of the major objections in their positions. Authors who later summarize positions often reflect a more doctrinaire, pedagogic approach that submerges many of the issues that have arisen in the course of the development of an approach. Thus, the approach in these essays depends on review of systemically developed and to that extent superior accounts of diverse positions. As indicated throughout, though, developments after these distinguished representatives have led to "living" or improving versions of these views.

Characteristic of such approaches are the lack of a back-and-forth methodology in which consequences force modifications in starting points and vice versa. More developed interpretations of each of these three representatives will indeed find such back-and-forth movements in their views, and these will become more evident in Chaps. 2, 3, and 4.

These three approaches may be called "Gaussian" to the extent that they hinge to a great extent on formal logic and/or regularity; the normal or Gaussian distribution, equilibria; and the assumption of the eventual funneling of results toward precise and unique solutions. That is, for any specific statistics (e.g., mean, standard deviation, 100-year loss), there is a real number toward which confidence intervals tend to converge to as the number of samples increase. Figure 1.2 illustrates this assumption in terms of various methods used to derive 95th centile confidence intervals. (Note, however, that even with the huge capacity and high speeds associated with the "cloud," Fig. 1.2 provides only a regularity that would vanish as the cloud's limits were exceeded.)

Figure 1.2 relates to Fig. 1.1 for fractile estimates as well as estimates for the mean estimates. In many cases (ignoring plateau issues), longer return intervals or lower probabilities converge much more slowly than more frequent return intervals or estimates of arithmetic mean values.

Although a later development, Bootstrap theory likewise emphasizes how in the long run estimates are necessarily made more certain in their gradual movement toward uniqueness.[13] Even for the deductivist theory, Carnap appears to feel most

---

[12] Some key works used by these authors include Carnap, Rudolf, 1962, *Logical Foundations of Probability*, Chicago: University of Chicago Press; Von Mises, Richard, 1957, *Probability, Statistics and Truth*, New York: Dover Publications, Inc.; and De Finetti, Bruno, 1970, *Theory of Probability: A critical introductory treatment*, Chichester West Sussex: John Wiley & Sons, Wiley Classics Library Edition published 1990.

[13] Efron, Bradley, and Robert J. Tibshirani, 1993, *An Introduction to the Bootstrap*, New York: Chapman & Hall.

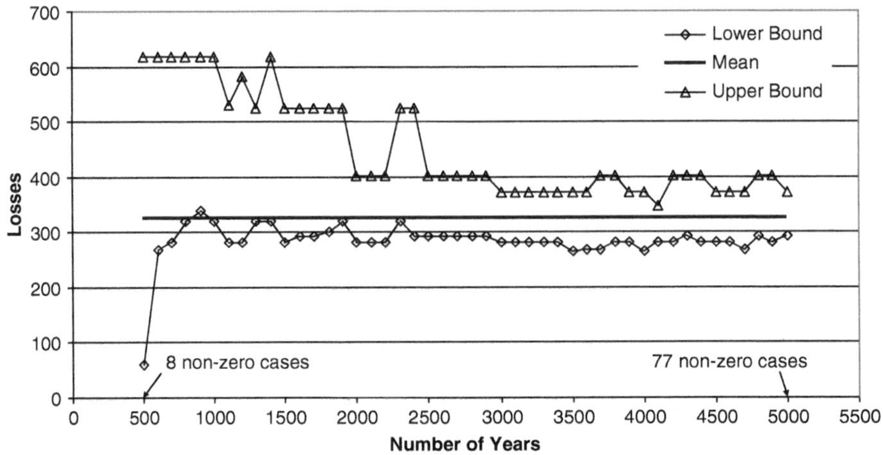

Fig. 1.2 Illustration as to how 95th centile confidence intervals converge to statistic of merit (here, the mean value for 5000 year samples, although other fractile estimates have similar results—ignoring Chap. 5 issues)

comfortable with mean values and Gaussian distributions, which were historically extraordinarily influential until late in the nineteenth century and remain prominent even today.[14] All of these three approaches have proven to have many valuable uses. Peter Bernstein remarks that the normal (Gaussian) distribution forms the core of most systems of risk management, although he later adds that at the extremes, the stock market, a risky place, is more likely to destroy fortunes than to create them.[15] Even the deductivist approach, which currently has few followers, provides enormous heuristic aids in learning and applying probabilistic and statistical terminology. Additional benefits throughout these essays of reviewing Carnap's approach include (a) noting the extent to which his foundationalist approach ignores all the successes of this binary logic in its offshoots in computer technology and (b) understanding how this binary logic can likewise impede understanding how major scientific and critical discoveries generally should especially in their original states be viewed as "nuanced" or "potentially promising" rather than either "T" or "F."

This book will emphasize how the three approaches arising from the eighteenth century have proven to encounter major challenges from research developments, events, processes, and applications emerging chiefly from the past century. These chapters will emphasize how long standing this belief in single solutions has been dominant yet how fragile has this opinion turned out to become.

[14] Von Plato, J., 1994, *Creating Modern Probability: Its Mathematics, Physics and Philosophy in Historical Perspective*, Cambridge: Cambridge University Press.

[15] See pp. 144 and 150 in Bernstein, Peter, op. cit.

A fourth ensemble of approaches to probability and statistics has emerged chiefly from this past century. Discussions of mega-risks, whether related to wars, financial breakdowns, and natural phenomena, include shocks to systems that cause what have been called "Black Swan" situations.[16] These have been defined as events or processes that are not very predictable, have very large-scale impacts, and often receive hindsight without foresight explanations. Black Swan issues cover not only these unpredictable or extremely random events but the failure to assure that the "next case" will be similar to those one has previously experienced. Journeys and expeditions to different regions, from the Galapagos Islands to Australia, from a city nearby to a rain forest, often illuminate how one's assumptions have been restricted to specific locales or habitats. Seeing houses built of brick, squirrels that are gray, and people who are emaciated may all reflect personal experiences. One may be surprised to find water systems that require heating for drinking or large-scale housing developments all with solar panels.

Shocks may be sudden as in the case of explosions or wildfires or may result from long-term natural or human activities, such as failure to dredge a river leading to a port, human and natural activities impacting changes in climate, or economic practices that eventually result in worldwide downturns. The more formal treatments of shocks in probability and statistics emerged largely in the 1920s and 1930s (and earlier for actuarial theories) and continue today.

Do the many theories developed only in the last century assist in understanding these kinds of Black Swan cases? These include theories of heavy-tailed (Levy) distributions, fractal theory (Mandelbrot), and pragmatic (inquiry-based, systemic, Dewey) theories.[17] Enormous advances in storage capacity and speed in information technology have provided advantages as well to digital approaches to probability and statistics.

So, to the three major approaches to probability and statistics, these essays combine a fourth:

> **The Pragmatic, systemic, inquiry-based approach**, that considers "robust simulation," a statistical approach that yields ranges of answers (in some or many cases, no unique solutions), along with other non-linear or systemic approaches including heavy-tailed distributions and fractal theory.

To regard this fourth viewpoint as a methodology in a narrow sense may be misleading. Robust simulation requires a process or strategy whereby alternative competitive evaluations are taken into account. Overall, the pragmatic viewpoint is

[16] See Taleb, N. N., 2007, *The Black Swan: The Impact of the Highly Improbable*, New York: Random House.

[17] For an updated account of Levy distributions, see, for instance, Nolan, John P., 2009, *Stable Distributions: Models for Heavy Tailed Data*, accessed on the Internet February 27, 2013; see also Mandelbrot, Benoit B., 1983, *The Fractal Geometry of Nature*, New York: W. H. Freeman and Company, originally 1977; see Dewey, John, 1938, *Logic: The theory of Inquiry*, New York: Holt, Rinehart and Winston, Inc., reprinted by Irvington Publishers, Inc. in 1982; for a pragmatic critique of other theories of probability and statistics, see Will, F. L, 1974, *Induction and Justification*, Ithaca: Cornell University Press.

consistent with incorporating results using fuzzy set theory, digital logic procedures, frequency tools, or Bayesian tools. And each of these three approaches (deductivist, frequency, or Bayesian) requires a high degree of coherence or self-consistency.

Oddly enough, in its early expressions, one finds in the pragmatist Charles Peirce a viewpoint that stresses convergence of opinions much like the convergence assumed in Bayesian analysis:

> On the one hand the followers of science are animated by a cheerful hope that the processes of invitation, if only pushed far enough, will give one certain solution to each question to which they apply it. [Different investigators] may at first obtain different results, but, as each perfects his method and his processes, the results are found to move steadily together toward a destined centre.[18]

The discussion of pragmatic theories generally will require addressing the challenging issue of how reasoning relies significantly on "consequences." Under some circumstances, the "fallacy of affirming the consequence" is genuine. In other cases, this so-called fallacy rests on the assumption of the linear view discussed briefly in the preceding.

This book hopes to clarify how robust simulation as well as nonlinear or systemic reasoning has arisen as a means of permitting the scientific and technical issues to be addressed in full detail by diverse investigators. From the standpoint of funding research, robust simulation requires competition as well as systematization. From the standpoint of those attempting to discover "the truth," the truth of "robust simulation" as it can be developed may best be characterized at any given time by an ensemble of competitive solutions. Systematic or nonlinear reasoning may on some occasions converge but in the process may allow considerable internal competition whether ultimate convergence occurs or not. Table 1.1 summarizes how the four main theories of probability and statistics fit into the following chapters.

The approaches in Chaps. 2, 3, and 4 are initially taken *categorically*, that is, as being true without condition. However, all three fail to be absolutely well founded. Although many angles exist from which these approaches may be criticized, the principal issue in these chapters resides in their "applicability."

Chapters 2, 3, and 4 may at first appear to result from a *via negativa* approach, an approach stressing "denial" of other positions. Karl Popper's "falsification" program represents one sort of via *negativa* approach.[19] These essays show how these updated eighteenth-century approaches do not account well for uncertainties in these mega-risk evaluations. In particular, major issues that arise in these theories, when taken categorically, are (a) null applications for the deductivist view and (b)

---

[18] From p. 38 in Peirce, Charles, 1878, "How to Make Our Ideas Clear," pp. 23–41 reprinted in Peirce, Charles 1955, *The Collected Writings of Peirce*, ed. Justus Buchler, New York: Dover Publications Inc. Interestingly enough, in 1896–1899, Charles Peirce writes that "experience can never result in absolute certainty, exactitude, necessity, or universality," (p. 47) and "there are three things to which we can never hope to attain by reasoning, namely, absolute certainty, absolute exactitude, absolute universality," on pp. 47 and 56 in "The Scientific Attitude and Fallibilism," pp. 42–59 in Charles Pierce, 1955, ibid. If convergence among diverse opinions were attainable, it would seem that we could arrive at some absolute certainty, exactitude, and universality.

[19] Popper, Karl, *1959*, *The Logic of Scientific Discovery*, *New York*: *Routledge Classics* 2002.

Table 1.1 Major theories of probability and statistics and some pertinent chapters

Chapter	Main term	Other terms	Proponents	Simplified description
Chapter 2	Deductivist	Logico-deductivist, logistic	Carnap	Deductive reasoning: logic and mathematics define statistics
Chapter 3, 6	Frequency	Inductivist, objectivist	von Mises, R. A. Fisher	More evidence results in more confidence; mathematical methods to achieve "fitting" and "significance"
Chapter 4	Bayesian	Subjectivist	de Finetti	May augment inductive inferences with personal or subjective prior estimates
Chapter 5, 7	Robust simulation	Pragmatic, systemic, inquiry based, nonlinear, outlier oriented	(Toulmin, many others)	A statistical approach that yields ranges of answers. Requires a process or strategy whereby alternative competitive evaluations are taken into account

null confirmations for the frequency and Bayesian theories, respectively. Following Chap. 4, the major issue becomes one of connecting experience and statistical mathematics.

However, the via *negativa* approach in these essays is far from a mere denial or falsification. Instead, affirmative views are here extracted from the rejection of aspects of thoroughly expounded positions. Imre Lakatos, a "follower" of Popper, maintains, "Criticism is not a Popperian quick kill, by refutation. Important criticism is always constructive: there is no refutation without a better theory."[20] That is, criticism does not suffice to stop with a purely negative attitude toward theories and the practice of probability and statistics.

1.3.2 Middle Chapters: The Dilemma of "Infinity"

Chapter 5 continues this apparent *via negativa* approach with respect to some very rare shocks or some rare systems responses to even minor shocks. Chapter 5 also addresses the possibility of "Black Swans," extreme value situations, or clear cases of non-convergence. The heavy dependence within Gaussian approaches on estimates of the mean and statistical variance is seriously questioned when extreme value distributions are involved. Statistical variance in its classical meaning no longer exists, and for even more extreme distributions, even the mean value in its

[20] From p. 6, Lakatos, I., 1978, *The methodology of scientific research programmes*, London: Cambridge University Press.

classical sense no longer exists. This chapter contains a discussion of catastrophe ("stability") measures and their relationship to power laws.

This chapter provides a simplified approach to extreme value theory, an approach merely designed to calibrate the "stability" (in the Gaussian sense) of the distribution with which one is working. This calibration is "empirical" insofar as only the finite data (simulated or real) is used for this calibration. Power laws illustrate the connection between the proposed catastrophe indexes and the evaluation of actual systems. Highly "dangerous" distributions can be associated with systems vulnerable to rare but extremely damaging events.[21] Deductivist, frequency, and Bayesian theories have assumed stability in models used, in spite of formal developments beginning in the 1920s that cover floods and other disasters and show that there are realities not consumed by regularities as assumed in the past. Figure 1.2 does not necessarily apply for the development of many extreme value distributions and for many statistics of interest (especially standard deviation, 100-year loss, and the like and in very extreme cases, the arithmetic mean and median).

Power laws are a simplified means to show how estimating risk to systems can be cognitively dangerous if only because the systems themselves as studied can be very "fragile" or have characteristics that render their behavior very unpredictable. So, while it may be obvious that in science various "shocks" may be very challenging to predict, it may also be the case that nonlinear behavior in systems can make evaluation of risk to these systems cognitively challenging, if not "dangerous."

Chapter 5 poses more problems with conventional views of statistics and so does not yet provide the framework for a constructive outlook. Bridging the mathematics or quantitative side of probability and statistics with experience becomes at this stage the major issue—an issue that had been raised historically by the early David Hume and later by J. M. Keynes.[22]

This issue may (accurately or not) be expressed "how does one bridge the gap between the infinite population assumed by probability and statistics with the finite samples of experience?"

Or, "how can one be sure that the next case or cases do not—as happens for extreme value distributions—significantly disturb one's current findings?"

To begin to address these bridging issues, the following chapter has been developed: Chap. 6, Mathematization of Statistics: Flexibility and Non-convergence. Chapter 6 provides an account of mathematical approaches that have been used in "fitting" models to distributions and in providing statistical "acceptability" tests of

---

[21] Beard, R. E., T. Pentikainen, and E. Pesonen, 1984, *Risk Theory*, 3rd edition, London: Chapman & Hall.

[22] General references are Hume, David, 1739, op. cit., and Keynes, and John Maynard, 1921, *A Treatise on Probability*, London: MacMillan and Co. Peter L. Bernstein, op. cit., p. 118, finds a quotation in Keynes, ibid., p. 368, ultimately from Leibniz, in which he regards contingencies in nature as prohibiting one from deriving life expectancy data based on very large data samples alone.

hypotheses. Although these activities had begun at least by 1900, the key figure used to begin this discussion is R. A. Fisher. One of his major critics today turns out to be the famous statistician and bettor, N. Silver.[23]

On the downside, the mathematization of probability and statistics has given rise to a huge inventory of "axioms," "theorems," "lemmas," "subroutines," "tests," and the like. This propensity to quantize science and other inquiries has a much longer history in the West than formal statistics and probability. N. Silver's critique of Fisher is that Fisher's reliance on the binomial, Gaussian, and Poisson distributions entails that he assumes knowledge of the total distribution that may not work in future cases and sequences of data. That is, Fisher assumes that there is but a single answer, that is, a convergence answer, when one is "fitting" data to distributions and developing significance tests for acceptability of hypotheses.

On the upside, though, this mathematization now provides a large number of tools that can be used to assess "fits" in diverse ways and to examine the "acceptability" of hypotheses from different vantage points. One does not need to be confined to 3 underlying distributions or to 30 or 40 currently common distributions: mathematicians have devised ways to construct an indefinitely large number of distributions that could be used in "fitting" and "testing." The flexibility of mathematics enables it to provide a cornucopia of tools that can be used in the evaluation of "fits" and "hypotheses." Chapter 6 provides potential future research in how some of the many possible mathematical tools may impact statistics on "fitting," "significance," and the like. In particular, if one cannot know with certainty in advance the trends of future data, how can one provide a more robust account of the diverse ways in which finite data samples are evaluated?

From the earliest chapters, it has become apparent that non-convergence in probability and statistics is much more common than has been assumed. Chapter 4, for instance, illustrates how non-convergent results may arise from very slight changes in a Bayesian prior. Chapter 5 shows how a "wobble" can be discovered in extreme distributions, a permanent oscillation that prohibits convergence for at least many statistics of merit. Chapter 6 shows that one can use mathematics to construct alternative "fits" and "significance tests." Chapter 7 continues these findings with first a discussion of how there are alternative credible models.

For many socially important activities, the author and colleagues have defined robust simulation as first requiring multiple outcomes. Having multiple outcomes by themselves does not overcome the issues raised in the previous chapters. In particular, these multiple outcomes cannot merely be the result of constructing a large number of alternative views that may or may not be attributed to someone. A mathematical search program alone with its ensemble of outcomes does not suffice to show how quantitative models are tied to experience.

---

[23] General references are Fisher, Ronald Aylmer, 1944, *Statistical methods for research workers*, London: Oliver and Boyd Ltd., ninth edition, and Silver, Nate, 2012, *the signal and the noise: why so many predictions fail—but some don't*, New York: the Penguin Press.

1.3.3 Robust Simulation: New Ways of Thinking

In Chap. 7, "Robust simulation" is defined.[24] "Robust simulation" is first discussed for more qualitative or social science activities, along with betting as well as minor Bayesian estimates. Robust simulation is next discussed in the evaluation of risks to complex systems. Multiple models are valuable to comprehend the "uncertainty" in estimates. Calculation of "confidence intervals" does not achieve this goal and the "uncertain" bounds of estimates are not themselves confidence intervals. Individual models may use Bayesian, bootstrap, frequency-based, fuzzy set, or other professional approaches. Rather than endorsing an ill-defined "subjectivity," "intelligence," addressing problems and borrowing on the rich cognitive resources available, is required to define the mental portion of successful investigations using finite samples of data. This intelligence is an active participation in the quest for knowledge through the application of pertinent disciplines. In effect, the professional community in its broadest sense must be involved in what amounts to a competition that illuminates the diverse trajectories of investigations.

No one of the professionally competitive multiple models in robust simulation constitutes "the truth." The presence of such state-of-the-art competition implies that at present, and maybe for a hundred or thousands of years, a singular truth may not be revealed. Nonetheless, on the view being advanced in these essays, these competitive results comprise knowledge or the state of the art. For robust simulation, a *process* must be in place that enables investigators to go in different directions that they approach a major issues or set of issues. Each investigative team may employ one or more diverse methods as they undertake the evaluation of risks of shocks to systems.

Thus, robust simulation inherits the simple catastrophe indexes in Chap. 5 and takes advantage of findings in the deductivist, frequency, and Bayesian modeling and "fitting" and "significance" tests in Chap. 6—as long as these do not require unique solutions for many risk evaluations of great social importance. These activities assist in developing the nexus between the quantitative side of statistics and experience, which also requires the long-term disciplined activities of the professional community.

Chapter 7 uses work by statistician David Freedman to further the discussion through case studies not only very briefly of major experiments that have taken

---

[24] Pertinent versions of "robust simulation" (and also catastrophe indexes) are found in Taylor, Craig, Yajie Lee, William Graf, Zhenghui Hu, and Charles Huyck, 2010, "Robust Simulation and Cat Diagnostics for Treating Uncertainties in Catastrophe Risk Analysis," pp. 155–163, in *Reliability Engineering and Risk Management*: *Proceedings of the International Symposium on Reliability Engineering and Risk Management*, ed. by Jie Li, Yan-Gang Zhao, Jianbing Chen, and Yongbo Peng, Shanghai, China, Tongji University Press, and Murnane, R. J., C. E. Taylor, T. Jagger, and Z. Hu, 2011, "Robust simulation for sensitivity analysis of catastrophe risk losses," in *Applications of Statistics and Probability in Civil engineering*, ed. by M. H. Faber, J. Koehler, and K. Nishijima, CRC Press, New York, PP. 875–877.

place but principally of major successes in addressing health system issues. Chapter 7 discusses how these major social issues (e.g., as in health systems) can be addressed through highly nonlinear statistical and qualitative approaches that eventually become part of a legacy of health system developments. These major health system issues were not solved definitively all at once, but the "successes" have become so because eventual developments have overcome initial objections and have further provided much more information on the nature and extent of these health system issues and way to manage and treat them when they arise. Dealing with human and other organisms (animal, bacteria), even the most complete health system discoveries mentioned, namely, the eradication of smallpox, have not reach an absolute lawlike state, one that has eliminated all future vigilance.

Characteristic of these very famous cases is the incompleteness combined with potential or fertility of the initial discovery. Typically, this initial discovery is followed by a very large number of developments in order to ascertain and define the details of the discovery. In effect, in many such cases, the initial discovery would be deemed "false" if taken by itself and if one ignored all the "consequences" that follow from follow-up investigations and applications. To evaluate such discoveries too precipitously would deprive them of their enormous value in providing insights into critical and practical activities to pursue in order to limit and validate them. These discussions assist in defining the expression "consequences," which are often after the fact developments of an initial idea that is not ripe for being simply called true or false. In their initial stages, many great discoveries provide viewpoints that compete with currently widespread beliefs. And because the development of ideas can take considerable time and effort, during the period of this development there is a strong possible that alternative and competing ideas will coexist.

1.3.4 Final Chapters: Possible Futures

Chapter 8 thus asks: "how does one provide a quantitative account of decision analysis for ensemble statistical outcomes?" Previously there have been stochastic accounts that have been provided using the principle of least regret, mean values, and statistical variances; the entire distribution of gains and losses (stochastic dominance), almost stochastic dominance, and fat-tail reduction models; and even the use of multiple decision criteria.

This chapter outlines some of these quantitative decision methods and hypothesizes as to how these may be used to assist in decision-making using robust simulation methods. This chapter concludes with the view that there are a number of quantitative tools that may help to make decisions based on robust simulation outcomes.

Chapter 9 covers remaining questions. In the elaboration of previous chapters, the treatment of various traditional approaches as being "categorical" is in some respects unfair. Treating viewpoints as being categorical facilitates an ultimately uncritical via *negativa* approach—an indefinite and pointless deconstruction. Yet the goal of Chap. 9 is to provide a plausible equilibrium in which for the time being the value of probability and statistics in dealing with systems is clarified. The presence of further queries implies that, even when one achieves some temporary equilibrium, there are many further lines of inquiry to be undertaken or in some cases to require assembly of what is there in various places already. Of special interest in Chap. 9 pertains to how the views of "robust simulation," "instabilities in extreme value distributions," and linear reasoning, among others, upset a very long-standing Western tradition of believing that there is but a unique solution, a singular truth to be achieved.

These multiple "interpretations" of mega-risks do not imply that one is endorsing a certain sort of very unreasonable vagueness and indefiniteness. Instead, what these multiple interpretations show is how alterative investigators can provide an ensemble of definite outcomes. Feigning that there is always but one outcome is to endorse a false precision rather than clarify and distinctness. The stochastic interpretations of complex phenomena clarify the range of current outcomes.

The use of robust simulation and the flexibility in mathematical and critical approaches to the use of probability and statistics both illustrate how our world view is enhanced by considering how alternative credible approaches and views are helpful and required in understanding a great many phenomena, including how to assess mega-risks.

A deep Western tradition assuming single solutions in critical studies—traced by Stephen Toulmin as deriving itself from developments in the seventeenth century—is one reason why many may not be able to accept the changes that come with such developments, as well as other developments that upset the ready-made world of statistics and probability theory. These developments go well beyond those mentioned in these essays. Chapter 9 points out, however, that findings from this ready-made world, when qualified, can be most useful going forward. The modest skepticism espoused by Toulmin and others should characterize the reasonableness within applications of probability and statistics as these continue to morph.[25] The blinding clarity of Newton, as commemorated by Alexander Pope, becomes a very

---

[25] Toulmin, Stephen, 1992, *Cosmopolis: The Hidden Agenda of Modernity*, Chicago: University of Chicago Press; a critique of the doctrinaire aspects of the Newtonian world view is found in Burtt, E. A., 1954, *The Metaphysical Foundations of Modern Science*, Garden City, N. Y.: Doubleday & Company, Inc., Doubleday Anchor Books.

important past glimmer of light as one moves through Plato's cave and regains one's vision with many subsequent significant vantage points.[26,27]

In this chapter, hypotheses are developed to treat these questions. Heavy emphasis is placed on how critical subjects are developed systemically. This systemic background contains idealizations and heuristics essential to the body of work completed or envisioned. This system background is thus itself subject to modifications and underscores the modifications implied by robust ensemble outcomes in important risk evaluations.

References

Beard, R. E., Pentikainen, T., & Pesonen, E. (1984). *Risk theory* (3rd ed.). London: Chapman & Hall.

Bernstein, P. L. (1996). *Against the Gods: The remarkable story of risk*. New York: Wiley.

Bruner, R. F., & Carr, S. D. (2007). *The Panic of 1907: Lessons learning from the market's perfect storm*. Hoboken: Wiley.

Burtt, E. A. (1954). *The metaphysical foundations of modern science*. Garden City: Doubleday & Company, Inc., Doubleday Anchor Books.

[26] In James Jeans, 1910, "On Non-Newtonian Mechanical Systems, and Planck's theory of Radiation," *Phil. Mag*., *20*, 943–954, we read.

> The question discussed in the present paper…is in brief as follows:-Can any system of physical laws expressible in terms of continuous motion (or of mathematical laws expressible in terms of differential equations) be constructed such that a system of matter and aether tends to a final state in which Planck's law is obeyed? It will be found that the answer is in the negative.

This is quoted on p. 204 in Thomas S. Kuhn, 1978, *Black-Body theory and the Quantum Discontinuity, 1894–1912*, Chicago: the University of Chicago Press. Although the major changes in hard sciences in the twentieth century are well celebrated, histories of the nineteenth century may show how research was adding to and modifying Newton's theory (even if one ignores his contemporaries such as Leibniz). For instance, an early history by J. H. Merz defines Newton as having constructed a mathematical physics with the fewest possible assumptions. Yet, many investigations in the nineteenth century led to new insights and medications of this view of nature with respect to fluid dynamics, compressibility, rigidity, mobility, elasticity, and electromagnetic fields and their lines of force. Merz also emphasizes very large changes as the focus in physics is on energy rather than force (pp. 27, 29, 40, 45, 59, 68, 69, 87, 145, and 198 in Merz, Theodore, 1903, *A History of European Scientific Thought in the Nineteenth Century*, Volume II, New York: Dover Publications, reprinted in 1965). On p. 132 in *The Concept of Energy Simply Explained*, New York: Dover Publications, Inc., Morton Mott-Smith furthers this by contending that revolutions in science come only when the system is already "tottering," and the doctrine of the conservation of energy was long overdue.

In The Feynman Lectures on Physics, Volume I: Mainly Mechanics, Radiation, *and Heat*, Richard Feynman maintains that since the 1920s, many forces have been recognized and the ultimate basis for forces consists of electromagnetic fields (2–3). Moreover, such chemical phenomena as nuclear reactions play a significant role, for instance, in astronomy (3–7).

[27] These questions and the stress on processes result from an adumbration of questions raised by M. Dresler in a review of the third draft of these essays.

Carnap, R. (1962). *Logical foundations of probability*. Chicago: University of Chicago Press.
David, H. (1968). *A treatise of human nature: Being an attempt to introduce the experimental method of reasoning into moral subjects*. Oxford: Clarendon Press.
De Finetti, B. (1970). *Theory of probability: A critical introductory treatment*. Chichester: Wiley, Wiley Classics Library.
Dewey, J. (1938). *Logic: The theory of Inquiry*. New York: Holt, Rinehart and Winston.
Efron, B., & Tibshirani, R. J. (1993). *An introduction to the bootstrap*. New York: Chapman & Hall.
Fisher, R. A. (1944). *Statistical methods for research workers* (9th ed.). London: Oliver and Boyd Ltd.
Friedman, T. L. (2006). *The world is flat*.
Gelman, A., Carlin, J. B., Stern, H. S., & Rubin, D. B. (2003). *Bayesian data analysis*. Boca Raton: Chapman & Hall/CRC.
Gladwell, M. (2002). *The tipping point*. New York: Little, Brown and Company.
Huebner, S. S., Black, K., Jr., & Cline, R. S. (1982). *Property and liability insurance* (3rd ed.). Englewood Cliffs: Prentice-Hall.
Hume, D. (1955). *An inquiry concerning human understanding*. Indianapolis: The Bobbs-Merrill Company, The Library of Liberal Arts.
Jeans, J. (1910). On non-Newtonian mechanical systems, and Planck's theory of radiation. *Philosophical Magazine, 20*, 943–954.
Keynes, J. M. (1921). *A treatise on probability*. London: MacMillan and Co. Peter L. Bernstein.
Kuhn, T. S. (1978). *Black-body theory and the quantum discontinuity*. Chicago: The University of Chicago Press.
Lakatos, I. (1978). *The methodology of scientific research programmes*. London: Cambridge University Press.
Murnane, R. J., Taylor, C. E., Jagger, T., & Hu, Z. (2011). Robust simulation for sensitivity analysis of catastrophe risk losses. In M. H. Faber, J. Köhler, & K. Nishijima (Eds.), *Applications of statistics and probability in civil engineering* (pp. 875–877). New York: CRC Press.
Nolan, J. P. (2009). *Stable distributions: Models for heavy tailed data*. Boston: Birkhäuser.
Peirce, C. (1878). How to make our ideas clear. Popular Science Monthly (January), New York, pp. 286–302.
Pope, A. (1954). The Poetry of Pope. In *EPITAPH, intended for Sir Isaac Newton* (p. 9). New York: Appleton-Century Crofts.
Popper, K. (2002). *The logic of scientific discovery*. New York: Routledge Classics.
Press, S. J. (1989). *Bayesian statistics: Principles, models, and applications*. New York: Wiley.
Shakespeare, W. circa. (1595). The merchant of Venice. In G. B. Harrison (Ed.), *Shakespeare: The complete works*. New York: Harcourt, Brace Jovanovich.
Silver, N. (2012). *The signal and the noise: Why so many predictions fail—But some don't*. New York: The Penguin Press.
Smith, N. K. (1966). *The philosophy of David Hume: A critical study of its origins and central doctrines*. New York: St. Martin's Press.
Taleb, N. N. (2007). *The black swan: The impact of the highly improbable*. New York: Random House.
Taylor, C., Lee, Y., Graf, W., Hu, Z., & Huyck, C. (2010). Robust simulation and cat diagnostics for treating uncertainties in catastrophe risk analysis. In J. Li, Y-G. Zhao, J. Chen, & Y. Peng (Eds.), *Reliability engineering and risk management: Proceedings of the international symposium on reliability engineering and risk management* (pp. 155–163). Shanghai: Tongji University Press.
Toulmin, S. (1992). *Cosmopolis: The hidden agenda of modernity*. Chicago: University of Chicago Press.
Von Mises, R. (1957). *Probability, statistics and truth*. New York: Dover Publications.
Von Plato, J. (1994). *Creating modern probability: Its mathematics, physics and philosophy in historical perspective*. Cambridge: Cambridge University Press.
Yajie, L., Graf, W., Hu, Z., & Huyck, C. (2010). Robust simulation and cat diagnostics for treating uncertainties in catastrophe risk analysis. In L. Jie, Y.-G. Zhao, J. Chen, & Y. Peng (Eds.), *Reliability engineering and risk management: Proceedings of the international symposium on reliability engineering and risk management* (pp. 155–163). Shanghai: Tongji University Press.

Part II
Discussion and Analysis of 17th-Early 20th Approaches to Understanding Risk

Chapter 2
The Deductivist Theory of Probability and Statistics

Cultural xenophobia is a frequent sequel to a society's decline from cultural vigor. Someone has aptly called self-imposed isolation a fortress mentality. Armstrong describes it as a shift from faith in logos, *reason, with its future-oriented spirit, "always...seeking to know more and to extend...areas of competence and control of the environment," to* mythos, *meaning conservatism that looks backward to fundamentalist beliefs for guidance and a worldview. A fortress or fundamentalist mentality not only shuts itself off from dynamic influences originating outside but also, as a side effect, ceases influencing the outside world.* (Jacobs, Jane, 2004, *The Dark Age Ahead*, New York: Random House, p. 17)

Ideas in modern Russian [the Soviet Union] are machine-cut blocks coming in solid colors; the nuance is outlawed, the interval walled up, the curve grossly stepped. (From p. 243, by Vladimir Nabokov, *Pale Fire*, 1962, New York: Perigee Books)

Abstract This chapter outlines one of three common views of probability and statistics: the deductivist view, discussed through a key proponent, Rudolf Carnap. This theory suggests that logic is the basis of mathematics and hence of probability and statistics. It requires absolute certainty in intuitions and deductions, thus not allowing for flexibility and adjustment. The chapter ultimately concludes that the deductive theory fails because of lack of applications. Today the view has few proponents. However, the limitations of this theory point to what is needed: more viable ways of comprehending uses of digital logic, the nature of starting points of inquiry, and the roles of consequences, applications, and problem-solving within inquiry. Carnap's view also provides key incipient insights into how "context" or "conditions" need to play a key role in understanding statistics and probability.

C.E. Taylor, *Robust Simulation for Mega-Risks*, DOI 10.1007/978-3-319-19413-4_2

2.1 Introduction

This chapter outlines one of three common views of probability and statistics: the deductivist view (aka foundational, linear). The term "deductive" in this sense comes from formal logic and mathematics alone. This view asserts that logic is the basis of mathematics and hence of probability and statistics. There is little doubt that mathematics and information technologies play key roles in probability and statistics. How to express this role, though, remains to be examined.

One very distinguished representative of this view is Rudolf Carnap.[1] This view of probability and statistics was once a major view but today seems to have few supporters. The sometime ascription of this "deductivist" view to J. M. Keynes adds nothing since this ascription is highly controversial.[2]

Carnap's very systemic view shows how a "backward-looking" approach continues to flounder owing to its requirement of absolute certainty in intuitions and deductions. Instead of adjusting one's view in order to account for some very important matters (e.g., the existence of irrational numbers), the view discussed here retreats to its foundations that are incapable of reaching such goals. This foundational view can be used by a purist in order to undermine very great science and critical views as they are developed.

In spite of these drawbacks, it is because of the powerful offshoots of the binary formal and mathematical logic in terms of a variety of competing computer programs and approaches that one is now able to see, for instance, Fig. 1.2, and how (ignoring the limitations of infinite samples even on the "cloud") statistical confidence levels gradually converge to a single point on a loss distribution. The nonlinear reasoning that has gone into these electronic and mathematical logical techniques shows the immense values of the logical studies that Carnap and others were undertaking.

This chapter exposes the major weaknesses of this viewpoint, all centered on how the approach leads nowhere. Going nowhere is common in experimental programs that are developed to learn from failures. Going nowhere in this case pertains chiefly to the underlying view of consequences, applications, and problem-solving. This chapter along with later chapters assist in understanding more viable ways of comprehending uses of digital logic, the nature of starting points of inquiry, and the roles of consequences, applications, and problem-solving within inquiry. These modifications are desired to reassess how to escape from the severe problems that Carnap finds himself in his quest for a risk-free theory of probability and statistics.

---

[1] Two direct pertinent references from Carnap are Carnap, Rudolf, 1962, *Logical Foundations of Probability*, Chicago: University of Chicago Press, and Carnap, Rudolf and Richard C. Jeffrey, 1971, *Studies in Inductive Logic and Probability*, Volume I, Berkeley: University of California Press.

[2] From Wikipedia [7], "A Treatise on Probability," accessed June 5, 2013. Keynes argued at some points that probability is strictly a logical relationship between evidence and hypothesis (e.g., P(H|E) is "logical), a degree of partial implication. See also footnote 10. Keynes is "a persistent subjectivist" according to Richard von Mises, p. 94 in 1957, *Probability, Statistics and Truth*, New York: Dover Publications, Inc. Keynes's work on probability, namely, *A Treatise on Probability*, encompasses a great many views.

This chapter questions the view that self-evident foundations of logic and mathematics provide the absolute certainty that undergird both applications and accounts of uncertainty in probability and statistics.

A more technical analysis of Carnap's "L"-language is reserved for an addendum even though this discussion has potentially considerable interest.

Finally, this chapter summarizes lessons learned from the evaluation of Carnap's very thorough but failed account of how one uses deduction as the means to account for probability and statistics. Given the deep tradition within which Carnap's view resides, these lessons are more challenging to absorb than many others in later chapters. However, Carnap's view provides key incipient insights into how "context" or "conditions" need to play a key role in understanding statistics and probability. In addition, had Carnap and others modified their view of the centrality of logic and adopted the approach of computer programs and systems in stressing all the uses of these programs and systems and had they looked forward rather than backward, the resulting modified view should have received widespread acclaim of computer programs and systems today.

2.2 Fundamentalism: From Physics to Mathematics to Logic

2.2.1 Descartes

One of the major thinkers stressing self-evidence was of course Descartes, with his *Cogito ergo sum*. Only with some parallels with Descartes' first Meditation, the development in this chapter leads from a type of universal skepticism through classical physics and through mathematics to mathematical logic.

2.2.2 Newton

In the previous chapter, the often quoted poem by Alexander Pope raised Newton and his insights into an exalted status. The belief that physics and in particular Newtonian physics was indefatigable was not avowed by all in Newton's time but has long since been shown to be qualified through centuries of modifications and "revolutions" in physics itself as well as such allied disciplines as chemistry. This sort of fundamentalism had vanished by the time that mathematics in the late nineteenth and early twentieth century became the object of a fundamentalist outlook.[3]

[3]As an aside, one of many accounts of fundamentalism, in religion, is Armstrong, Karen, 2001, *The Battle for God: A History of Fundamentalism*, New York: Ballantine Books. Arguably, a similar type of zealous fundamentalism can surface in critical disciplines, as indicated in Burtt, E. A., 1954, *The Metaphysical Foundations of Modern Science*, Garden City, N. Y.: Doubleday & Company Inc., Doubleday Anchor Books. Burtt maintains that this zealous attitude prevailed by and toward Newton's work. An extension of this theme is found in Toulmin, Stephen, 1992, *Cosmopolis: The Hidden Agenda of Modernity*, Chicago: University of Chicago Press. Toulmin

Attempts to develop a "logicist" theory of mathematics, i.e., a view deriving mathematics from pure logic, became seriously questioned in the early twentieth century. Furthermore, diverse views of mathematics have arisen throughout its history. For example, according to Herman Weyl, there are two prominent views of mathematics: the constructivist and the hypothetico-deductivist. He favors the first view:

> By the mathematical way of thinking I mean first that form of reasoning through which mathematics permeates into the science of the external world—physics, chemistry, biology, economics, etc., and even into our everyday thoughts about human affairs. ... The power of science, as witnessed by the development of modern technology, rests upon the combination of a priori construction with systematic experience in the form of planned and reproducible reactions and their measurements.[4]

This view is chiefly a functional view, stressing functions, constants, and variables. This view binds itself to material interpretations. This view returns especially in Chap. 6 on the mathematization of probability and statistics, which can be treated as being "constructivist."

In contrast to this constructivist view, Weyl mentions the deductive method in which the mathematician is "left to himself" and less interested in truth than in consistency.[5] The chapter covers this second view of mathematics: the deductive method of developing probability and statistics.

So, if mathematics is not the discipline sin which self-evidence and deduction prevail and yield risk-free truths, what is left for those seeking risk-free knowledge? And the answer for many in the twentieth century, even such notable thinkers as Imre Lakatos, is mathematical logic. Because this faith in mathematical logic has extended so far and wide, this chapter spends considerable space on what feature of this answer: the attempt to treat probability and statistics as merely a derivative of mathematical logic.

2.3 Carnap

Carnap reflects the twentieth-century philosophy in America and Great Britain, which has been dominated by this view of the relatively isolated sphere of logic and mathematics, especially logic and its view of rigorous inquiry and expression. Risk-free logic is taught as being fundamental to all science and cannot be questioned within science itself. This Logic is not falsifiable and undergirds definitions of falsifiability.[6]

---

maintains that a doctrinaire attitude prevailed in the West from 1610 and for at least 300 years and associated with historical circumstances that create a heteronomy in scientific studies.

[4] From pp. 1832 and 1844 in Weyl, Herman, "Mathematical Creation," pp. 1832–1849 in *The World of Mathematics*, 1956, ed. by James R. Newman, New York: Simon and Schuster.

[5] From pp. 1832 and 1846 in Weyl, Herman, Ibid.

[6] References in this paragraph are to Carnap, 1962, op. cit., pp. 161, 577, and 244.

Carnap's purported goal is a reconstruction "restricted to a simple language form, of inductive thinking in everyday life and science…a critically corrected reconstruction." Carnap expects through his reconstructive quest to give mathematical statistics for the first time a solid foundation, a systematic unity of its various methods, and a clarity and exactness of its basic concepts.[7]

2.3.1 *The Problems of Applying Carnap's Viewpoint*

2.3.1.1 Illustrative Carnap-Like Axioms

Given that Carnap adopts a risk-free approach to the theory of statistics and probability, one might expect him to use an axiomatic approach and in particular one that begins with self-evident axioms and proceeds through deductions to self-evident consequences. One begins with the following[8]:

- There is a set E of elementary events x, y, z,…
- There is a family of subsets **F** of E. Its members are called *chance* events.

Derived from Kolmogorov, one might imagine that the following axioms are acceptable to a Carnap-like view:

1. As a field, **F** is closed with respect to unions, intersections, and compliments.
2. F contains E.
3. To each set A of **F**, a nonnegative real number P(A) is attached. This number P(A) is called the probability of the event A.
4. P(E) = 1.
5. If A and B are disjoint, P(AUB) = P(A) + P(B)

These require that some events are "elementary." These propositions further assume that probability statements are constant. Propositions can only be applied when one uses, for instance, sets or else "types" to categorize. Propositions are either true or false. The theorem of total probability obtains—that is, all of the probabilities for a given phenomenon must yield one. These basic axioms therefore appear to apply to languages in which the terms used are fixed and clear cut.

---

[7] See p. 52, Carnap, 1962, Ibid.

[8] This treatment is derived from Von Plato, J., 1994, *Creating Modern Probability: Its Mathematics, Physics and Philosophy in Historical Perspective*, Cambridge: Cambridge University Press. pp. 217–220. In reconstructing Carnap's position, Mather (2009) maintains that "K" can be defined as a "fixed" proposition and is called "background evidence." Notably, however, it is more than merely challenging to construct "K" as a clear-cut set. Note as well that the Bayesian approach to total evidence requires a finite partition and so implies that there is only a finite sample of evidence. References for Carnap, 1962, Ibid., are pp. 211, 212 and for Maher, Patrick, "Explication on Inductive Probability," in Formal Epistemology Workshop, June, accessed from the worldwide web on October 9, 2009.

As a caveat to the above, in his quest for explicitness, Carnap follows Keynes and others in stating probabilities as P(H|E), the probability that H is true given E for two sentences H and E. P(H) alone is not thus a basis for this pure probability theory. This P(H|E) has a unique and constant value.[9]

To render his axioms applicable to statistics and probability, Kolmogorov adds the following rules permitting the application of probability and statistics to experience[10]:

1. A certain complex S of unlimitedly repeatable conditions is assumed.
2. One investigates certain events which may appear in the realization of the conditions S. In individual cases of the conditions, the events appear in general in different ways. Let E be the set of possible variances x_1, x_2,... of how the events appear. The set E contains all variants we hold a priori as possible.
3. If the variant appearing after the realization of conditions S belongs to the set A, we say the event A appeared.
4. Under certain conditions ..., one can assume that to the event A a real number P(A) is attached such that:

   (a) If the conditions S are repeated a great number of times n, one can be *practically certain* that the relative frequency m/n of occurrence of A differs only a little from P(A)
   (b) If P(A) is very small, one can be practically certain that A does not appear in a single realization of the conditions S.

The extension of the original mathematical axioms of probability theory to these rules of applicability poses many obvious problems, especially with respect to discussions of catastrophes. The following discussion does not assume that Kolmogorov thought that these rules of applicability were risk-free. Carnap himself did not regard such rules as being risk-free. For Carnap, existing relative frequencies may be upset by future evidence. Black Swans—when the future is not like the past—are possible.[11]

Proposition 4a ignores such dramatic changes and trends that can upset what one estimates as a probability. If for many years there have been many trees in the forest, does the probability of there being at least two trees in the forest depend only on this isolated experience? Or, does it depend on, for instance, such processes as deforestation and recent fires? Landscapes can change rapidly.

Proposition 4b ignores preparing for mega-risks. Probabilities of individual mega-risks tend to be very low. However, practice requires not that one turns aside from such potential disasters but considers them in light not only of their probabilities but also in terms of the severity of their consequences. Think of ignoring the potential for tsunamis in Asia and of ignoring the potential for storm surge in New Orleans and the practical consequences of this negligence.

---

[9] From Carnap, 1962, op. cit., p. 37. The general reference to Keynes is to Keynes, John Maynard, 1921, *A Treatise on Probability*, London: MacMillan and Co. Whether or not Keynes is a pure logicist is moot.

[10] From Von Plato, J., 1994, op.cit.

[11] This is presumably the reason for the title in Taleb, N. N., 2007, *The Black Swan: The Impact of the Highly Improbable*, New York: Random House.

For catastrophes, we may be evaluating the "law of small numbers," that is, regularities in catastrophic or rare events.[12] Are, we may ask, hurricane frequencies regular, or have increases in ocean temperature increased their intensity and hence their frequency of occurrence? Or, can we assume that earthquakes have in a given region a stable energy output over time based on a small sample of very large earthquakes (which tend to dominate such statistics since the increase in one magnitude level corresponds to about a 31-fold increase in energy output)? If this is so, then very rare magnitudes of 8.0 or above greatly impact the "average" energy (or moment) produced in a given region of the world. Magnitudes of 9.0 and above have an even more dramatic impact on earthquake statistics.

2.3.1.2 Carnap Himself Admits that His Efforts Lead to a Null Set of Applications

As indicated in his text, on the one hand, one might expect Carnap to be arguing that one can justify his logic in terms of its success, and this might include a lengthy discussion of how information systems have evolved from logic. On the other hand, Carnap maintains that logic is what it is, apart from its applications, and that it should be universal for any system of concepts that fit the particular language in question. Put in other terms, Carnap's logic should not commit the "fallacy of consequence" but should instead justify matters in a linear fashion—one step at a time.[13]

In constructing this system, Carnap begins with a simple language L that has only a finite or denumerably many entities. Carnap then goes on to say that this system fails because it cannot account for the continuums found in science: space, time, temperature, mass, length, and so on. As has been found in many sources, one needs to have real numbers in order to account for these continuous values, and Carnap's simple L-language has no such capability. Similarly, without the use of space, time, and other continuous values, one cannot distinguish individuals from each other. Thus, Carnap himself admits that his "logic" has yet to be so extended in a "self-evident" fashion if such an extension were even feasible. Carnap's quest as of the 1950s failed to have applications in science. As shown later in this chapter, Carnap in the 1960s resorts to a theory of betting in order to try to achieve applications.

2.3.1.3 Carnap Creates a Separate Sphere of Logic and Mathematics: A Risk-Free System of Logic and Its Extension to the Theory of Probability and Statistics

On the deductivist treatment of probability and statistics, logic and mathematics are set apart and are logically prior to any scientific, engineering, financial, social science, or any other cognitive activities. Logic and mathematics are distinct from

[12] See Keynes, John Maynard, 1921, *A Treatise on Probability*, London: MacMillan and Co., pp. 403ff.

[13] References to the views favoring applicability are in Carnap, 1962, op. cit., pp. 7, 108, and 161.

methodology. In this separate sphere of logic and mathematics, one can achieve logical and mathematical truths that are true in any universe. "Probability" can be defined in this separate sphere, and so "probability" can become exclusively a realm of logical or a priori truth. Deductions alone exist in this risk-free sphere.

This deductivist sphere is not designed to expand on the tools that logic and mathematics have in order to solve a variety of older and new problems. Certainty rather than expansion of possibilities and applications is the goal of this deductivist view.

Noticeable in such an extreme view as Carnap's and his followers is the use of many rhetorical flourishes or defense mechanisms to assure that their views—expressing risk-free truths—combat any assaults from outsiders. Those who stress "real life" suffer from "abstractophobia"—defined by Carnap as "depriving science of some of its most fruitful methods." Introverts, who stress abstractions rather than observations, have been vital to science. Those who treat logic as critical thinking are subject to "psychologism." Applications of logic lie outside logic and, not being the task of logic, are inessential to its truth.[14] One should not raise questions about proposed elementary and universal truths. One should merely intuit their truth. Elementary truths should be expressed in an axiom system. Derivations too should be completely transparent. Precision and exactness are to replace vagueness. "Applications" are not "pure."

The creation of a risk-free sphere has serious consequences for any attempt to validate mathematical logic as through its consequences, especially but not exclusively in such consequences as its value in the development of computers. In the 1980s, R. Feynman developed his lectures on computation in which he began with the sort of binary logic in mathematical logic. One may cavil that computer binary logic is not completely identical with mathematical logic: for computers, not everything follows from a contradiction. However, more importantly, in the development of a computer, Feynman goes on to explain how, for instance, communication (electronic) theory must be deployed and the thermodynamics of computing must be considered. In both cases, very small errors can arise in the machine and reducing the probability of errors must be considered in light of other factors salient in the development of a computer. In short, in order to *apply* logic through the use of a computer, one cannot eliminate all risks. The situation is of course worse when human beings are involved in the application of a "pure" logic.[15]

2.3.2 The Basic Critique of This Foundationalist View: The Road to Nowhere Argument

For all foundationalist views like the deductivist account, there is an argument that maintains that if one achieves "self-evidence," then one has achieved, so to speak, a road to nowhere. No further consequences can be derived from a view that is by

[14] From Carnap, 1962, Ibid., pp. 208, 216, and 217.

[15] See Feynman, Richard P., 1996, *Feynman Lectures on Computation*, edited by Tony Hey and Robin W. Allen, Cambridge, MA: Perseus Publishing.

itself so certain that no further evidence or derivations can falsify or modify it. Very briefly speaking, any consequences that one may derive from a self-evident view can only create cognitive trouble for this purportedly self-evident view. Cognitive trouble for a view deemed self-evident means that the view is not as self-evident as it was initially deemed. First impressions of this truth are upended by second impressions when this initial truth has consequences that may be unwanted.[16]

This basic critique has been shown to apply with respect to an axiom system that does not yield applications to probability and statistics, to the discovery that the real number system cannot be developed from Carnap's logic, and to the failure to verify the logical system used in terms of such enormous applications as those in the development of computers, from the failure to understand that errors, however slight, occur in applications including verification that logic has been used.

The finding that the foundationalist or linear view does not account even for the continuum or real numbers, let alone complex numbers, is very damaging to Carnap's viewpoint. This finding results from an account of the binary account of the fallacy of affirming the consequence as follows:

> It is a fallacy to maintain that "If p implies q, then if q is true then p is true" in which "p" and "q" are each propositions that are either T or F.

The similar fallacy within the theory of self-evidence is as follows:

> It is a fallacy to maintain that "p is self-evident" when p implies q and q may be false. Thus, if p implies q, and p is self-evident, then q must be true.

The above "fallacies" help to clarify the reciprocity of antecedent and consequent when expressed in binary terms. As later chapters will indicate, these "fallacies" stated in binary language do not reflect nuanced views of truth and falsity. In Chap. 7, examples will be given of how major discoveries are often expressed in incomplete ways, and the "consequences" of these major discoveries may take generations to clarify. One such example comes from F. Waismann's work on mathematical thinking:

> …the differential quotient is not a quotient at all, but the limiting value of a sequence of quotients. However, this was not yet clear to the founders of differential calculus, although they occasionally came very close to the truth. By and large, they cultivated the view that the differential quotient is the ratio of the quantities Δ x, Δ y at the instant at which they just vanish—the *ultima evanescentium incrementorum*, as Newton said. Leibniz and Newton had a feeling that there existed a difficulty in the formation of this concept; however, they were unable to get a really clear idea about it.[17]

---

[16] The view that first impressions, including those in mathematics, can be overridden or modified by later developments is found in the psychological work of Kahneman, Daniel, 2011, *Thinking, Fast and Slow*, New York: Farrar, Straus and Giroux. Slow thinking can overcome failures in fast thinking. One treatment of the philosophy of mathematics that shows how those studying mathematics or even doing arithmetic undergo transformations as they proceed from natural numbers to integers and so on is found in Waismann, Friedrich, 1951, *Introduction to Mathematical Thinking*, New York: Frederick Ungar Publishing Co. A much more extended "road to nowhere" argument is found in Taylor, Craig Elliot, 1974, *An Essay on the Possibility of Inference*, Ph.D. dissertation under Professor Frederick L. Will, Champaign, IL: University of Illinois (unpublished).

[17] See p. 150, Waismann, Friedrich, ibid.

We can now use and compute with infinitesimals after a long period of confusion. That confusion, while needing to be removed, did not prevent the progress of the subject. This volume will continue in the tradition of working with incomplete logical foundations in hopes that later workers will reinforce and solidify them.[18]

Thus, when the immensely useful notion of the differential quotient was introduced, its brilliant founders did not have a clear idea about it and it took time for this idea to become clarified. So, if the starting point of an inquiry should be true in an unqualified sense, then the differential calculus started from a very shaky foundation owing to the lack of clarity of the notion of a differential quotient.

As Feynman has indicated, information technology requires taking into account very small errors in the electronics of computers. In addition, digital computation itself has great challenges relative ultimately to "analog" issues involving real and complex number systems. The presence of limitations does not imply that enormous value comes from such digital electronic systems. The failure of Carnap's approach to account for the value of real and complex number systems (implied in Kolmogorov's axioms) and digital electronic systems is the basis for the stringency of the "road to nowhere" line of reasoning. The role of logic, mathematics, and information technology in probability and statistics is undermined through this foundationalist approach.

Frequency theorists as we shall see postulate that the future evidence will yield a long-term forecast or convergence to some value P(H). Frequency theorists thus define probability differently not as something logical but as something in which, so to say, the future will be like the past.

2.3.3 Selected Anachronisms in Carnap's View

To identify some selected anachronisms in Carnap's view is merely to identify how modifications in probability and statistics, and related areas, have had consequences that would require changes in Carnap's view. Two of special interest are first Carnap's preference for using the mean value and Bernoulli processes and second Carnap's attempt to use an embryonic view of decision theory.

2.3.3.1 Carnap's Preference for an Estimate of the Mean Value μ

Carnap distinguishes between probability in the logical sense (probability_1) versus probability in the nonlogical sense or frequency (probability_2).[19] No certainty exists that the estimate of μ is equal or even near to the actual value μ. This again assumes

---

[18] Dr. Robert Riehemann, letter dated December 22, 2014.

[19] There is further a search for c*, the quantitative explicatum for probability_1, a representative of the concept of degree of confirmation (p. ix). Inductive logic, in its quantitative form, may be regarded as the theory of c. In selecting a primary candidate for c, Carnap picks P(H|E) = the esti-

that there is some actual value μ. Still, Carnap maintains that various proposed principles used to derive μ directly—rather than as an estimate—fail. These include the doctrine of the uniformity of nature, namely, that the future will be like the past.[20]

In speaking of the reliability of estimates of μ, the e-function or confirmation function, Carnap appears to rely on extensions of Bernoulli processes in which the confidence intervals are derived based on such a formula as

$$\mathrm{P(H \mid E)} + / - [t * \sigma / (n - 1)] \tag{2.1}$$

in which t is derived from a table for the normal distribution and σ is the standard deviation derived relative to H given E. n is number of samples. If there are distributions such as the Cauchy that have no finite mean, there are even more distributions including the Cauchy that have no finite variance or standard deviation. In addition, especially for finite samples, there are many distributions such as the exponential distribution in which the confidence intervals so derived are not symmetric, in contrast to what the foregoing formula implies. Thus, Eq. (2.1) does not fit the discussion later in Chap. 5 about extreme value distributions. There are as well many ways to develop confidence intervals when there is a finite variance or standard deviation.[21]

The selection of the arithmetic mean and the use of the binomial distribution as principal factors are curious. Later in Chap. 6 we witness the statistician/biologist R. A. Fisher likewise resorting to fairly simple statistics and distributions. However, as seen in Chap. 5, if the underlying distribution were a Cauchy distribution, then there is no finite mean. Other very "heavy-tailed" distributions—those with a small percentage of very high absolute values—may lack a stable estimate of the mean.[22]

---

mate of the mean value for H, given only E. Once H and E are given, the probability ip(H,E) is logically derived and fixed forever. This probability is the estimate of the relative frequency. If presumably there were no relative frequency μ and no probability$_2$, then there would be no probability$_1$.

[20] References in this paragraph are from Carnap, 1962, Ibid., pp. 169 and 178–180.

[21] For supporting references, see Hogg and Klugman, 1984 and Law and Kelton, 1991, and also see Carnap, 1962, Ibid., pp. 510, 512, 534, 536, 537, 564, and 582. Alternative methods for developing confidence intervals given finite variances are found in treatments of the Chebyshev inequality, bootstrap resampling methods, and, when they apply, the use of control functions (whether or not combined with bootstrap methods). For bootstrap methods, see Efron, Bradley, and Robert J. Tibshirani, 1993, *An Introduction to the Bootstrap*, New York: Chapman & Hall. If Carnap has selected the most cognitively certain method for estimating confidence intervals, he would have selected Chebyshev's inequality, which is analytically derived given a finite variance (see pp. 141–142 in Meyer, Paul L., 1970, *Introductory Probability and Statistical Applications*, Reading, MA: Addison-Wesley Publishing Company).

[22] The Cauchy distribution is mentioned merely in passing by Carnap, 1962, Ibid., p. 245. This distribution, of course, creates huge challenges for a theory of probability and statistics that always assumes that the mean value μ is finite. In Triana, Pablo, 2009, *Lecturing Birds on Flying: Can Mathematical Theories Destroy The Financial Markets*? Hoboken, New Jersey: John Wiley & Sons, Inc., the author posits that financial markets are so wild that a Cauchy distribution may be the best one to apply since this distribution is so extreme and has only a constant median value. Not all extreme value distributions have even a constant median value.

The other comments in this paragraph are explicated later in Chap. 5.

2.3.4 Carnap's Later Excursions into Decision Theory

In 1971, Carnap considers problems of "applications" of a "pure" (and "universal") logic. At this later time, he still considers the reliance on "frequency" theory to be unreliable because "statistical probability values" are not generally known.[23] Instead, he defines in this context "probability" to mean "degree of belief." One begins with a belief system that permits one to evaluate alternative bets (or acts) in terms of outcomes with certain probabilities in the world. The bridge to "applications" does not reside in these probabilities, since they are not known. Instead, they are part of a belief system, and this belief system must be at least self-consistent internally and the same as the belief system of other "bettors." This betting procedure supplies Carnap with his bridge to applications.

This bridge results from our alleged ability to determine exactly and uniquely that our bets cannot be improved. Our logical betting system corresponds to the betting system of the "House," that cannot go into ruin but that cannot beat our logical betting system. So, our bets are suitably logical and probabilistic if we cannot lose to a superior bettor.

Very often, the advantage of the "House" or other bettors results from superior information. So, the supposition here must be that our systems of beliefs about acts, outcomes, and probabilities must be identical. So, how have quantitative decision procedures fared since their development in the early half of the twentieth century?

Carnap is placing his bets on a unique solution for a "maximum" bet. However, as developments before 1971 and definitely afterward have shown, there is no unique solution for each and every self-consistent betting system.

In the 1950s, Markowitz and others developed a "mean-variance" approach to financial or betting decisions. Two dimensions are used to evaluate investments or bets: the mean return on investment and its variance. Investments are superior to the extent that the mean return is higher and the variance is lower. Thus, an alternative with the highest mean or expected return on investment may also have a very high variance and so be less desirable than an alternative with the same mean return but a lower variance. The use of two dimensions for evaluating investments or bets assures that there may be many alternative betting schemes that are "indifferent" to each other in the sense that they could be acceptable to someone at some time. As has been shown many times, the use of a unique "utility" function to define a unique solution has many flaws.[24]

A more advanced approach that fits Carnap's desire for a "maximum" betting value comes from work by H. Levy and others who develop "stochastic dominance" theory. First-order stochastic dominance occurs when one alternative is better or at

---

[23] See Carnap, Rudolf and Richard C. Jeffrey, 1971, op. cit., pp. 8– 9.

[24] References for this paragraph include Markowitz, H. M., 1959, *Portfolio Selection: Efficient Diversification of Investments*, Oxford: Basil Blackwell Ltd., and Kahneman, Daniel, 2011, *Thinking, Fast and Slow*, New York: Farrar, Straus and Giroux.

least no worse for each and every possible situation. This, though, is a very rare situation and not typically useful in investment procedures. Second-order stochastic dominance does ensure that the betting system with the highest mean return can never be rejected. However, this conclusion still permits other betting systems to be in the same situation. The uniqueness of the betting solution thus vanishes with this more advanced system of betting, much as it does with the mean-variance approach.

Note that "almost stochastic dominance" may prove to be a more valuable tool inasmuch as it downplays small differences between alternative betting schemes and stresses major differences between these schemes. Note also that none of these betting schemes fully addresses circumstances in which first, second, and even third moments are infinite. The embryonic decision procedures that Carnap was relying on do not render probability applications "a priori," "purely logical," or known indisputably.[25]

2.3.5 Summary of the Deductivist Theory

The Upsides

- Carnap's rigorous quest leads him to reject common postulates such as the uniformity of nature, namely, that future cases will mirror past cases. There can be black swans even if one has only seen white swans. There are often unforeseen consequences when one expects and acts on past regularities.
- Carnap recognizes that real numbers are immensely challenging if not impossible to derive (through "efficient procedures"). This is even more true of geometric reasoning, including not only Euclidian and non-Euclidian geometries but nowadays fractile geometry.[26] The extensive use of digital computers raises similar issues as does the use of finite samples in statistics.
- Carnap recognizes that probability and statistics are conditional. This conditionality begins as a precise P(H|E) and devolves into a vague P(H|E&K), in which K becomes very vague. However, conditionality of probability and statistics is extremely important in following chapters.[27]
- Carnap's emphasis on a binary logic has strong parallels (and some nonparallels) with the digital developments in information technology (IT). IT advances have

[25] Stochastic dominance has been comprehensively outlined by Levy, H., 2006, *Stochastic Dominance: Investment Decision Making Under Uncertainty*, 2nd edition, New York, NY: Springer. Applications to natural hazards events have been developed by Taylor, Craig, Glenn Rix, and Fang Liu, 2009, "Exploring Financial Decision-Making Approaches for Use in Earthquake Risk Decision Processes for Ports," *Journal of Infrastructure Systems*, Volume 15, Number 4, pp. 406–416, December 1, 2009.

[26] See Mandelbrot, Benoit B., 1983, *The Fractal Geometry of Nature*, New York: W. H. Freeman and Company, originally 1977.

[27] Maher, Patrick, 2010, "Explication on Inductive Probability," in Formal Epistemology Workshop, June, retrieved from the World Wide Web on October 9, 2009.

greatly enhanced the ability of users to publicize and use a large number of tools in probability and statistics.

The Downsides

- Carnap attempts to maintain that mathematics is founded on logic and that one can devise a logic of probability. He hopes that logic, mathematics, and probability can be restricted to a risk-free sphere, a sphere of the analytically true, the a priori truths. This quest, expressed in Carnap's metalanguage, is not without its risks, and the language that Carnap devises in the *Logical Foundations of Probability* fails miserably in this work. There are minor inconsistencies in this quest, such as when P(H|E) is analytic, and both H and E are said to be atomic or logically independent. There are major challenges when Carnap confines himself to equiprobable distributions and the stress on mean values. Save in cases in which population statistics are known, underlying distributions may indicate trends, clustering, or other memory-based events. There are evolutionary reasons why the British saw only white swans until the late seventeenth or early eighteenth centuries. Likewise, mean values may be misleading or in some cases incalculable.
- Major inconsistencies arise in Carnap's quest when the language L as devised has no applications to become universal when the quest is furthered to include real (and perhaps complex) numbers. Whereas the quest for such foundations may have invaluable offshoots, this quest tends to preclude any potentially precarious applications or risky metalanguage interpretations that exhibit foundations that shake and slump. Any quest to construct an a priori sphere of truth separate from applications suffers from lack of adequate feedback loops. The variety of computer programs that have emerged have resulted from competitive settings in which users find advantages and disadvantages of various systems and select those best suited for their uses. A priori logic treats feedback consequences that are in any way negative as being instead signs of "incompetence."
- By ignoring consequences of his theory, Carnap ignores validations of the same theory. By ignoring applications and the errors that enter even into computer applications of a theory, Carnap places the theory in an isolated zone to which one has no access. The recognition that digital information technologies pose small mathematical and physical limitations on processing findings in probability and statistics does not imply that these information technologies should be epistemologically abandoned.
- Carnap implicitly rules out a number of applications in science and elsewhere. Most notably, many authors have maintained that species are not as clear when one views evolutionary history. Black swans may originally have been white swans but, after reproductive isolation, may have morphed and become a separate species.[28] Transitions states, mutations, and a variety of other anomalies arise in a more rigorous and hence less clear-cut evaluation of evolutionary phenomena. Species are clearly not points in space time and are hardly separable

[28] Wikipedia, "Black Swan," accessed April 28, 2013.

from their habitats. Evolutionary biology and language may transcend the axioms of Carnap or Kolmogorov and their assumptions of primitive individuals, elementary events, sets or types, and atomic propositions.

- Likewise, the pursuit of how the earth's surface has changed is much more complex and indirect than what Carnap's account of atomic propositions permits. Since linguistic changes are common in the development of continental drift theories (e.g., "crustal plates"), fixing language in advance, whether in terms used or the number of terms used, lacks adequate flexibility. The evidentiary basis for such estimates as the temperature of the earth's interior and many other phenomena in continental drift theory exhibit the use of highly refined instrumentations and complex calculation routines and assumptions to derive even single data points.[29]
- The goal of this deductivist approach is not to expand how mathematics can be used to illuminate problems in critical subject matters. The goal is to define a risk-free approach to probability and statistics, and in this regard it is left with an approach that cannot be used: a road to nowhere.

Addendum 1: Carnap's "L"-Language with Only Denumerable Values and His Admitted Limitations for Scientific Applications

As indicated in the text, on the one hand, one might expect Carnap to be arguing that one can justify his logic in terms of its success, and this might include a lengthy discussion of how information systems have evolved from logic. In contrast, Carnap maintains that logic is what it is, apart from its applications, and that it should be universal for any system of concepts that fit the particular language in question. Put in other terms, Carnap's logic forbids that should not commit the "fallacy of consequence" but should instead justify matters in a linear fashion—one step at a time.

In *Logical Foundations of Probability*, Carnap presents two languages: L_n and L_∞. The former consists of a finite system and the latter consists of a denumerably large system. In general, one must form the system from

- Countably many logically independent individuals, i_1, i_2, …
- Clear-cut, independent, and binary (either applicable or not) primitive classes/predicates/properties/modalities (presumably color, shape, and so on)
- Atomic sentences—which ascribe primitive properties to an individual
- Extensional combinations of atomic sentences (for everything complex or "molecular," one uses an inclusive "or," a tilde for negation, or an "and")
- Binary assignments ("T" or "F") to all sentences so formed.

[29] Uyeda, Seiya, 1978, *The New View of the Earth: Moving Continents and Moving Oceans*, San Francisco: W. H. Freeman and Company, pp. 2, 11ff.

Finite or denumerably infinite systems are basically in terms of information technology "digital" as opposed to "analog." Only a countable number of variables, sentences, individuals, and state descriptions exist. In this digital world, irrational numbers become truncated. Statistical distributions become a disjunction of a countable number of individual distributions. So, how does one apply such a language L (whether L_n or L_∞)? If individuals are logically independent and primitive properties are logically independent, how can probabilistic statements be logical? Suppose that a probabilistic statement is of the Carnap/Keynes formulation P(H|E) or the probability of H given E. Suppose further that both H and E are atomic sentences. Then, the sentences H and E therefore imply some value ip(H|E), a conditional probabilistic estimate. To that extent, H and E are not independent. H and E are thus relevant to each other or else ip(H|E) = 0, still a logical implication. Thus, at a minimum, there are serious challenges for deriving probabilistic estimates for two atomic sentences in Carnap's theory of probability.

Digital information systems—along with many related products such as digital pictures—has had enormous successes—but with some obvious sacrifices. To what world does this digital logic apply without sacrifices? Carnap's response appears to be that there is no world that he can think of that fits languages L_n or L_∞ perfectly. He states:

> It seems best to imagine as individuals in a system L, not extended regions like physical bodies or events in our actual world, but rather positions like the space-time points in our actual world, hence unextended, indivisible entities. Since, however, the number of individuals in a system L is either finite or denumerably infinite, they cannot form a continuum… the qualities and relations with which we are acquainted in our actual world cannot, strictly speaking be applied. For instance, a color occurs in the actual world only as a property of extended, continuous area.

Later, Carnap rules out length, mass, and temperature. One might hope that a statement of the form "at point (x,y,z) at time t the temperature is T," as being suitably atomic, but not even this is the case. Carnap admits that the actual language of science and even that of elementary physics has, of course, a much more complex structure.[30] The whole language of science has "its great complexities, its large variety of forms of expression, and its variables of higher levels (e.g., for real numbers)." This, though, requires a much more thorough rational reconstruction than Carnap pursues in his *Logical Foundations of Probability*. The simpler languages so far constructed have no applications whatsoever, and so the null set of applications is all that remains at this point of this "universal" logic.[31,32]

The status of the metalanguage may be reconsidered at this stage. As with other proposed languages and their elements, the metalanguage is clearly not an L-language. The metalanguage is one may presume supra-logical, that is, beyond

---

[30] From Carnap, 1962, Ibid., pp. 73–74.

[31] See pp. 541, 208, and 209 in Carnap, 1962, Ibid.

[32] See pp. 199 and 163 in Carnap, 1962, Ibid.

the restrictions of the L-language logic. Statements about the author's goals—while to be taken at face value—as well as statements about other's misconceptions stretch well beyond the sphere of risk-free logic.

Note that with sacrifices or tradeoffs, this logic could have immense applications and hence successes. Alternatively, Carnap in his later writings and his followers may pursue a course of a risk-free logic that is also universal.

References

Armstrong, K. (2001). *The battle for God: A history of fundamentalism*. New York: Ballantine Books.

Burtt, E. A. (1954). *The metaphysical foundations of modern science*. Garden City: Doubleday & Company Inc., Doubleday Anchor Books.

Carnap, R. (1962). *Logical foundations of probability*. Chicago: University of Chicago Press.

Carnap, R., & Jeffrey, R. C. (1971). *Studies in inductive logic and probability* (Vol. I). Berkeley: University of California Press.

Efron, B., & Tibshirani, R. J. (1993). *An introduction to the bootstrap*. New York: Chapman & Hall.

Feynman, R. P. (1996). *Feynman lectures on computation*. Cambridge, MA: Perseus Publishing.

Jacobs, J. (2004). *The dark age ahead*. New York: Random House.

Kahneman, D. (2011). *Thinking, fast and slow*. New York: Farrar, Straus and Giroux.

Keynes, J. M. (1921). *A treatise on probability*. London: MacMillan.

Levy, H. (2006). *Stochastic dominance: Investment decision making under uncertainty* (2nd ed.). New York: Springer.

Maher, P. (2010). Explication on inductive probability. *Journal of Philosophical Logic, 39*, 593–616.

Mandelbrot, B. B. (1983). *The fractal geometry of nature*. New York: W. H. Freeman and Company.

Markowitz, H. M. (1959). *Portfolio selection: Efficient diversification of investments*. Oxford: Basil Blackwell Ltd. and Kahneman, D. (2011) *Thinking, fast and slow*. New York: Farrar, Straus and Giroux.

Meyer, P. L. (1970). *Introductory probability and statistical applications*. Reading: Addison-Wesley Publishing Company.

Nabokov, V. (1962). *Pale fire*. New York: Perigee Books.

Taleb, N. N. (2007). *The black swan: The impact of the highly improbable*. New York: Random House.

Taylor, C. E. (1974). *An essay on the possibility of inference*. Ph.D. dissertation under Professor Frederick L. Will, Champaign: University of Illinois (unpublished).

Taylor, C., Rix, G., & Liu, F. (2009). Exploring financial decision-making approaches for use in earthquake risk decision processes for ports. *Journal of Infrastructure Systems, 15*(4), pp. 406–416.

Toulmin, S. (1992). *Cosmopolis: The hidden agenda of modernity*. Chicago: University of Chicago Press.

Triana, P. (2009). *Lecturing birds on flying: Can mathematical theories destroy the financial markets?* Hoboken: Wiley.

Uyeda, S. (1978). *The new view of the earth: Moving continents and moving oceans*. San Francisco: W. H. Freeman and Company.

von Mises, R. (1957). *Probability, statistics and truth*. New York: Dover Publications, Inc. Keynes's work on probability, namely, *A Treatise on Probability*.

Von Plato, J. (1994). *Creating modern probability: Its mathematics, physics and philosophy in historical perspective*. Cambridge: Cambridge University Press.
Waismann, F. (1951). *Introduction to mathematical thinking*. New York: Frederick Ungar Publishing Co.
Weyl, H. (1956). The mathematical way of thinking. In J. R. Newman (Ed.), *The world of mathematics III* (pp. 1832–1849). New York: Simon and Schuster.

Chapter 3
The Frequency Theory of Probability

This book is about a new, fourth paradigm for science based on data-intensive computing. In such scientific research, we are at a stage of development that is analogous to when the printing press was invented (p. xiii in Bell, Gordon, 2009, "Foreword," pp. xiii–xvii in Hey, Tony, Stewart Tansley, and Kristin Tolle, ed, The Fourth Paradigm: Data-Intensive Scientific Discovery, *Redmond, Washington: Microsoft Research).*

Learning to use a "computer" of this scale may be challenging. But the opportunity is great: The new availability of huge amounts of data, along with the statistical tools to crunch these numbers, offers a whole new way of understanding the world. Correlation supersedes causation, and science can advance even without coherent models, unified theories, or really any mechanistic explanation at all. There's no reason to cling to our old ways. It's time to ask: What can science learn from Google? (These deliberately provocative words are from Chris Anderson, 2008, "The End of Theory: The Data Deluge Makes the Scientific Method Obsolete," Wired, *6/23/08.)*

As Bellerophon's fame grew, so did his hubris. *Bellerophon felt that because of his victory over the Chimera he deserved to fly to Mount Olympus, the realm of the gods. However, this presumption angered Zeus and he sent a gad-fly to sting the horse causing Bellerophon to fall all the way back to Earth (From "Bellerophon," Wikipedia, the free encyclopedia, accessed 10/17/2013).*

Abstract One of the most popular theories of probability and statistics is the "frequency theory" originating in the eighteenth century. This data-driven view is developed here through a chief proponent, Richard von Mises. Von Mises is concerned in a major way that laws of large numbers and the central limit theorem follow from any account of probability and statistics. He defines a "collective" containing unrelated or fairly inconsequential data. He thus does not capture trends, correlations,

C.E. Taylor, *Robust Simulation for Mega-Risks*, DOI 10.1007/978-3-319-19413-4_3

perturbations, and a host of phenomena from the realm of statistics. Frequency theory also ignores extreme values or outliers. The chapter ultimately concludes that the theory assumes a convergence at infinity that will never be experienced. However, frequency theory vies with the Bayesian theory in current-day popularity because of claimed successes.

3.1 Introduction

This chapter discusses one of the most popular theories of probability and statistics: the "frequency theory" with origins in the eighteenth century. This view is developed here largely through a chief proponent, Richard von Mises. The frequency theory is one of many tools of interest when the right tail of the loss distribution is small (however long).[1] The frequency theory places enormous emphasis on data and its evaluation, with important parallels with the data-mining approach now in full swing.

While making initially huge concessions to the uses of frequency theory (and Bayesian theory) relative to light-tailed distributions, this chapter will not address the well-known small sample issue. The doctor may need few tests to determine that the patient's blood pressure is in a dangerous zone; the structural engineer may deem that a building suffers from soft first story problems in a high seismic zone. In many cases, few trials are needed to arrive at a very reliable result. This chapter covers only large sample issues.[2]

The frequency theory can provide enormous assistance in addressing the original questions raised in these chapters. It is conceded throughout this chapter that many loss distributions are light tailed. However, its serious limitations suggest that it is not to be taken as the entire answer. The version of this theory provided by von Mises provides enormous insights into the difficulties of using laws of large numbers for the vast majority of data-mining activities, especially those pertaining to shocks to systems. Both frequency theory and data mining, though, greatly enhance our knowledge through the quest itself.

[1]A very helpful approach to efficiencies is found in Chap. 23 of Efron, Bradley, and Robert J. Tibshirani, 1993, *An Introduction tothe Bootstrap*, New York: Chapman & Hall. In general, the topic is one of variance reduction techniques, and control functions provide potentially very great simulation efficiency gains for light-tailed distributions. One need not combine these with the very useful bootstrap modeling to produce, for instance, confidence intervals that can account for mean [fractile] estimates as well as say 5th and 95th centile estimates.

[2]On pages vii, 156, 158, 159, and 163 in *Probability, Statistics and Truth*, New York: Dover Publications, Inc., 1957, Richard von Mises opposes small sample theory including Bayesian theory and the use of case studies.

3.2 Von Mises: Search for a Simple and Exact Theory

The frequency theory is discussed here through the major figure Richard von Mises.[3] In sharp contrast to Carnap, von Mises maintains both that the most important test of probability theory lies in its applicability and that probability theory provides an experimental (natural science) basis for probability statements. Probability theory is concerned with forecasting and not merely with providing "descriptive statistics." Like Carnap, von Mises maintains that one must begin with the simple and exact solution and extend and improve it gradually.[4]

Linear at the outset, the theory of von Mises requires developments in statistics that are prior to its application. In particular, his account of a "collective" requires that one curate data so that it definitely permits the application of the law(s) of large numbers. This curation thus requires a back-and-forth statistical movement not characteristic of perfectly linear approaches.

This discussion of von Mises will begin with his definition of "probability" and then proceed to his notion of "collective" that he regards as being critical to his notion then to his discussion of the law(s) of large numbers. He regards his notion of the collective as critical to an adequate account of these laws.[5] Afterward, an illustration of large sample statistics is used to emphasize the conditioning and conditionality of the uses of these statistics.

For von Mises, probability is defined in terms of relative frequency. For instance, if the temperature in San Francisco exceeds 90 °F 5 days a year, then its relative frequency is 5/365.25 = 20/1,461. Because von Mises seeks a simple and exact solution that accounts for probability and statistics, probability and statistics apply to mass phenomena in which the same event repeats itself again and again, or a great number of uniform elements are involved at the same time. Thus, one needs years of data to develop an adequate relative frequency for temperatures in San Francisco. Population statistics covering birth and death rates; social statistics covering marriages, suicides, crimes, incomes, and heredity; medical statistics covering the action of drugs and cures; economics covering mass production, consumption, prices, demand, lotteries, gambling banks, and life insurance companies; and many studies in physics (such as those of Brownian motion) comprise activities in which such mass phenomena are studied.[6]

In particular, the *relative frequency* is "*the ratio of the number of cases in which the attribute has been found to the total number of observations.*" *Probability is the constant limiting value of this long* [*theoretically infinitely long*] *sequence of experiments.* That is, any probabilistic statement concerning temperatures in San Francisco

---

[3] From Von Mises, Richard, 1957, ibid.; See Von Plato, J., 1994, *Creating Modern Probability: Its Mathematics, Physics and Philosophy in Historical Perspective*, Cambridge: Cambridge University Press, pp. 13ff.; According to Keynes, 1921, op. cit., pp. 92ff.

[4] From von Mises, 1957, ibid., pp. 8, 30, 53, 54, 100, and 166.

[5] From von. Mises, 1957, ibid., pp. 80, 113, and 125.

[6] From von Mises, 1957, Ibid., pp. 11, 18, 102, and135.

is validated by a long number of years of thermometer readings. The resulting probability is unique. That is, there is a real number that corresponds to a given estimate of the probability that 90° is exceeded on a given day in San Francisco. This real number is not less than zero and not greater than one. For von Mises as for Carnap, probabilities are unique.[7]

To assure that he begins with a simple and exact theory, Von Mises contends that there is no "gaming system" (Von Mises's term for a "system of selection"). The mass phenomena must be completely random or lawless. *The observation of one sample does not impact the observation of the next.* For instance, if one measures the inflow of water at a given junction on a water trunk line and the outflow at a second junction, with no other outlets along this line, one would expect there to be a connection between the inputs and, after a time, the outputs. Of course there may be losses along the line and pressure and possibly temperature changes may create variations in this relation, but still one would expect the data from these two junctions to imply a statistical connection. The data together would not constitute a random sample in the sense that von Mises is using. One might try considering the set of data from each junction to be considered a part of separate collectives—if the data stream from each junction could be considered as being random in itself if the data samples were far enough apart in time so that each did not impact the next.

3.2.1 The Simple and Exact Solution: Search for a "Collective" from the Plethora of Data and Test for Equally Likely Random Samples

Some accounts of data mining as the first two quotations speak of data and its vast accumulation as a possible source of immense optimism in how rapidly knowledge and its use can benefit humanity. Gannon and Reed speak of "a rising tsunami of data" in environmental, healthcare, and biological disciplines that can move science from asking about data for hypothesis testing to asking about correlations, insights, and cross-disciplinary patterns.[8]

In serious data-mining activities, curation and similar activities require considerable effort and thought. "Provence data" have been defined as requiring the history of inputs and processing steps along with other information to assist in determining the relevance of these data to other activities and further to establish their veracity and support further activities involving reproducibility.[9] Putative data can be erroneous, duplicative, confusing, only marginally supported, and irrelevant.

[7] From von Mises, 1957, Ibid., pp. 14, 33, and 127.

[8] From pp. 131–132 in Gannon, Dennis and Dan Reed, 2009, "Parallelism and the Cloud," pp. 131–135 in Hey, Tony et al., op. cit.

[9] From Van De Sompel, Herbert and Carl Lagoze, 2009, "All Aboard: Toward a Machine-Friendly Scholarly Communication System," pp. 193–199 in Hey, Tony et al., op cit.

Von Mises undertakes his own curation tasks: to assure that samples are suitable for probability and statistics. The curation that von Mises seeks to undertake eliminates a huge part of the plethora of data so-called in data-mining activities. Thus, one may take von Mises as having a reductionist attitude toward the vast amounts of data that are available. One small subset, in effect, survives the curation that von Mises undertakes. Note again that von Mises undertakes these data reduction activities so that the laws of large numbers can obtain.

One chief concern for von Mises is that the previous deductivist theory treats probability as the ratio of the number of favorable cases to the total number of *equally likely cases*. Yet, this is very confusing. For von Mises, one must perform trials in order to assure that cases are equally likely. For instance, one may have as one datum the sighting of Haley's comet and another datum might be a car accident in San Bernardino County. These are not equally likely events. So, von Mises believes that one must test the trials to determine that they are equally likely.[10]

So, for von Mises, the central problem of statistics is finding out whether or not a certain group can be considered a collective. He thus tries various attempts to "reduce" sequences to those that can be treated according to a method much like sampling with replacement. For instance, if one draws a card from a deck of cards and then returns the card to the deck, then probabilities for drawing a 1, 2,... do not change on the next draw. Von Mises likewise tries to show how Brownian motion can be reduced to a "collective." At the same time, he contends that there are many mass phenomena to which the theory of probability does not apply.[11]

In order not to be encumbered by the mathematical conclusion that there is always a rule for any finite sequence, von Mises postulates that the "collective" from which one develops trials must be infinite. The sequence itself is thus assumed to be infinite.[12] Thus, drawing without replacement from a deck of cards in which there are only a finite number of possible draws does not count as a "collective."

For von Mises, to discover that there is equiprobability, one must assure that there is no statistical "memory," no dependence of the next case on previous cases. For instance, if only a finite number of possibilities are available, sampling without replacement yields dependence on previous cases. For example, if from a deck of cards one draws a 10 and does not replace it in the deck, then the denominator for a probability reduces to 51, and this modifies probabilities for drawing each of the cards, a 2, a 3, a 4, and so on. Thus, from a finite collection, samples without replacement comprise one example of lack of randomness for von Mises. *Equal likelihood for von Mises entails statistical independence. The collective must not contain samples drawn from a finite set and drawn without replacement. The collective must not contain samples whose probability is known in advance.*

---

[10] From Von Mises, 1957, op cit., pp. 53, 54, 67, and 80.

[11] See Von Mises, 1957, op. cit., pp. 141–145.

[12] See Von Mises, 1957, op. cit., pp. 91, 92, 93, and 101. In Chap. 4, the derivation of Bayes' theorem presupposes a finite partition of the universe in question. Chapter 4 brings out some problems with this finite sample when it is used for a large number of forecasts.

For von Mises, to be *random*, the sample in question must not be able to be impacted by some system of selection. One such nonrandom system of selection might be cyclical. One's trials must not be such that one can sample the 5th, 10th, 15th, and so on elements of a long sequence and so arrive ultimately at a different estimate of probability.[13] Thus, Haley's comet must be ruled out as a sample along with many astronomic phenomena such as lunar cycles. So, too, routine behavior such as going to school every weekday must be ruled out except when it ceases to be metronomic behavior. For von Mises, *equal probability is random, and both equal probability and randomness imply the absence of a system of selection as occurs when there is cyclic behavior. The collective must not contain samples that exhibit cyclic behavior. Again, probabilities of samples must not be known in advance.*

Similarly, trends must be discovered through trials. Blood pressure tests over time may indicate that a patient has increasing systolic and diastolic numbers. One needs empirically to discover whether or not the trials are equally likely, or whether or not they exhibit a trend. Global warming must be ruled out to the extent that it contains samples of ongoing diminutions of the polar ice cap, with consequences elsewhere. Conflagrations following natural hazard events may be less likely as cities impose building codes and land-use requirements that limit fire spread. The heavy use of DDT to control insects may lead to considerable loss of animals and plants. Deforestation without replacement of trees can yield losses of the wide variety of uses of timber as well as other impacts.[14]*For von Mises, randomness and equiprobability imply that there is no gaming system so that there are no trends. The collective must not contain samples that are part of trends.*

If trends, cycles, dependencies, and unequal probabilities are ruled out, what is left? For heuristic purposes, let us suppose that we draw n samples from a collection having infinitely many (denumerably many) elements or events. For instance, let us assume that there are a great many cards and that theoretically aces occur in 1 of 13 cards that are drawn. For von Mises, one cannot postulate that the probability that aces are drawn is 1 out of 13. Instead, one must draw the cards and derive the relative frequency from the draws. Some of the cards could stick together and so upset the theoretical or a priori probabilities. Now, one would expect each draw to be an event, and the actual card drawn (whether a 2, 3, …, or ace) to be a property or attribute of the event. So, one is concerned with the relative frequency of the attribute, given that the draws are equiprobable. Figure 3.1 shows how this statistical cycle (with an expectation of 1 in 13 draws being an ace) varies over 100 random draws. There is considerable difference in how long it takes before the ace is drawn. In some cases, after an ace is drawn, another ace is drawn next. In other cases, there may be many draws before an ace is drawn. Figure 3.1 assumes that whenever an ace is drawn, it is replaced in the deck. Thus, the draw is one with replacement.

---

[13] From Von Mises, 1957, Ibid., pp. 24, 25.

[14] On the use of DDT, see Carson, Rachel, 1962, *The Silent Spring*, Boston: Houghton Mifflin Company. On impacts of deforestation, see Diamond, Jared, 2005, *Collapse*: *How Societies Choose to Fail or Succeed*, New York: Viking.

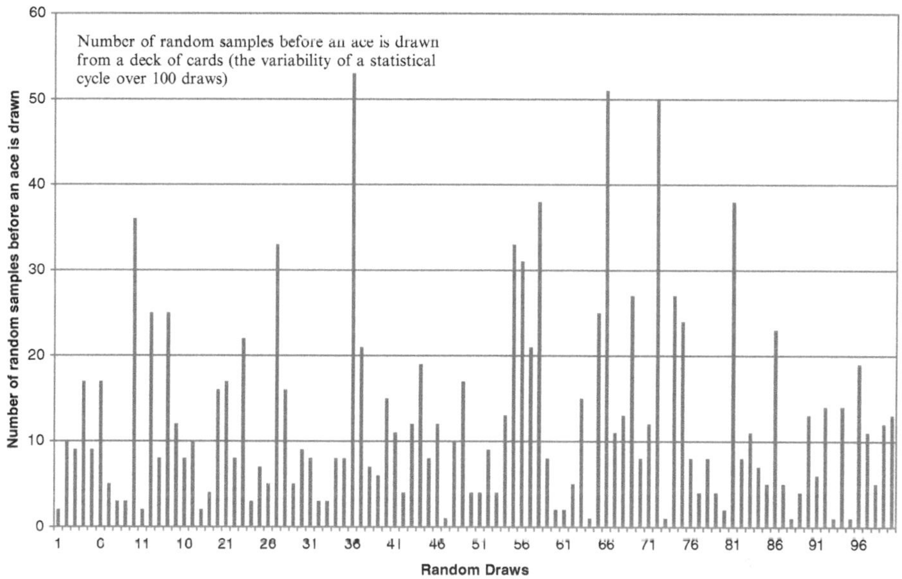

Fig. 3.1 Expected number of throws between aces in N throws of a die

In this way, one can imagine the process to be similar to a computer simulation that begins with 13 cards and uses uniform random numbers in order to draw cards. If the computer draws the cards without replacement, then of course the probabilities change with each draw. So, instead, one may try a computer program in which the computer draws cards with replacement. If the uniform random numbers selected are as random as possible, then one would expect aces to be drawn on the computer about 1 in 13 times in the long run. Thus, one can see something like Fig. 1.2 in how the computer in the long run converges toward the result.

In general, one might try a random selection of cases drawn with replacement in order to determine if this is adequate to fit von Mises's definition of a collective. For instance, suppose that one wants to pick jurors randomly. Assume that there is a compiled database from which one may derive people who are at least 18 years old, registered to vote, not felons, not defined as a "dependent" for reasons of dementia or otherwise, living in the suitable jurisdiction, and not having served in the past 3 years. From this list, one may use random numbers in order to select say 30 possible jurors to be evaluated by the prosecution and defense. This actually is a case of sampling without replacement, but at least there is a way in which issues pertaining to various biases may be minimized—subject to review by lawyers on the case in question.

However, this is not the way in which von Mises is viewing the matter. Sampling without replacement from a finite set is not his approach. Instead, one must view the situation as being one in which sampling is done from a potentially infinite set. Sampling with replacement constitutes a situation in which one may believe that one views the potential set as being infinite. Sampling without replacement from an infinite set does not mean that the probability of each draw is actually 1/n. Actually, each draw has in the collective a probability of $1/\infty$, and so sampling without replacement fails to meet von Mises's test.

For sampling with replacement, a different possible problem arises. If one draws enough sample cards, then there will be a "statistical" cycle that develops for aces. *On average*, aces will be drawn 1 in 13 times. This does not of course imply that there will be 1 ace in each 13 draws. In some cases of 13 draws, there will be 2 or 3 aces so drawn. In many other cases of 13 draws, there will be no aces drawn. Still, one expect there to be a statistical cyclic behavior that surfaces as the draws continue.

Sampling jurors nonetheless is less likely to be allowed. Still, using sampling with replacement, one might devise a rule that some prospective juror may be selected two or more times on the same jury. A cycle is removed only if (a) any juror no longer appears again on the list of prospective jurors, (b) the number of prospective jurors for a case changes, or (c) the size of the population from which jurors are selected changes. For both jurors and cards, removing statistical cycle behavior will require considerable effort—not very realistic in practice.

So, typically ruling out trends, dependencies, cycles, and the like, von Mises has greatly reduced the data that are suitable for use in a collective. The proportion of data in a von Mises "collective" is but an extremely small portion of all the data in today's data-mining activities. The curation process used by von Mises deflates the vast data-mining optimism so that he may achieve a more credible theory of probability and statistics. At the same time, since von Mises believes that only experience can assure that there are cycles, dependencies, trends, and other perturbations, then in order to reject various data samples from a collective, an enormous effort must be made that yields considerable information. So, it is only as regards uses of various statistical laws and the like that data-mining activities are reduced by the view promoted by von Mises. In ruling out various data samples from being part of a collective, there will be considerable statistical tools used.

Thus, a typical criticism of von Mises's view of the collective and its random or equiprobable samples is that one must have knowledge in order to tell that there is no "gaming system" or "system of selection." That is, experience must have played a role in evaluating the cyclic behavior of Haley's comet or the moon. There appears to be a vicious circularity in this definition of randomness.

This criticism is strengthened because collectives are assumed to contain infinitely many samples. As Keynes points out, definitions of randomness that rely on "the long run" fail because they require complete knowledge.[15] In effect, Keynes is maintaining that all of the references to one's previous success in finding randomness or in determining a final result in the "long run" are confused. *In no case have we experienced successfully an infinite sequence of trials.*[16]

[15] References on this paragraph are first from Von Mises, 1957, op. cit., p. 137 and then from Keynes, 1921, op. cit., p. 290.

[16] Elaborating on this idea, on p. 6 *Applied Chaos Theory: A Paradigm for Complexity*, San Diego CA: Academic Press. 1993, A. B Cambel states

Complete knowledge may lead to the conclusion that Haley's comet is more random than is expected, that the diminution of polar ice caps is more uneven than previously thought, and that not all communities suffer virtual extinction when they tolerate deforestation. Considerable thought and time, inferential power, is required in the curation of samples that von Mises is undertaking.

3.3 The View that the Law(s) of Large Numbers Require Collectives

In earlier or Gaussian stages of probability theory and statistics, the binomial theorem has served as a major means of confirmation of viewpoints. Von Mises rightly criticized the earlier versions of the law of large numbers for not first purging the samples of cases that may yield considerable errors. In this earlier version, this law is as follows:

> If an experiment, whose results are simple alternatives with the probability p for the positive result, is repeated n times, and if ε is an arbitrarily small number, then the probability that the number of positive results will be not smaller than n(p-ε) and not larger than n(p+ε) tends to 1 as n tends to infinity.[17]

> Because a set of numbers appears to be random does not mean that it is. Statistical analyses must be undertaken to establish the nature of the data. For example, ...in card games a deck must be shuffled seven times before the odds that a card may be in any position are the same. In turn, two decks must be shuffled nine times...

Further elaborating on this idea on p. 18, in *Randomness*, Cambridge, Massachusetts: Harvard University Press. 1998, Deborah J. Bennett maintains that

> Von Mises defined randomness in a sequence of observations in terms of the inability to devise a system to predict *where* in a sequence a particular observation will occur without prior knowledge of the sequence....Yet certainly every sequence conforms to *some* rule—we may simply not know what the rule is ahead of time....In a 1963 paper Andrei Kolmogorov was able to show that if only *simple* formulas, rules, or laws of production are allowed, then von Mises-type sequences would exist....[For Kolmogorov] a random sequence is one with maximal complexity...if the shortest formula which computes it is extremely long. ...The problem is, How do we ever know if we have found the *shortest* formula? ... Common to all of these views is the *unpredictability* of future events based on past events.

The reader who wishes to learn about random coin tossing should read [quoted in J. Ford, 1983, "How Random Is a Coin Toss?" *Physics Today*, April, pp. 40–47; note that the first known six-sided dice are dated as being from the East in 2750 B.C. See also von Mises, 1957, op. cit., pp. 69, 74, and 85 on how the manufacturer, toss and eventual wear and tear of the die will impact the results.

[17] From Von Mises, 1957, Ibid., p. 105. This statement has a close relationship with Eq. 2.1 that Carnap uses in order to gain some applications of probability and statistics.

To show how nonrandom samples can upset this rule, von Mises constructs a sequence of 0s and 1s that does not obey this theorem in terms of the intervals assumed, even though the probability of 0s ultimately ends up as p. For von Mises, the appropriate law must be empirical. Unlike the stated law above, von Mises further asserts that one should let the data speak rather than presupposing that p is the probability in question.[18]

This approach has problems for small, large, and huge computer programs that postulate probabilities in advance. For instance, one may say that the computer program that includes cards picked at random has a bias if only the conventional 13 (bridge, poker, etc.) cards are selected, and a quasi-random (as random as possible) method is used for selection. For von Mises, should there be such biases?

For estimating probability p, von Mises restates the law of large numbers as follows:

> The ratio of numbers derived from the observation of a very large number of similar events remain practically constant, provided that these events are governed partly by constant factors and partly by variable factors whose variations do not cause a systematic change in a definite direction.[19]

Variable factors that cause a systematic change in a definite direction may include trends, causes, systemic effects, correlations, cycles, and so on. As von Mises knows, small causes may have large effects. So, collectives that are ruled out include samples of smallpox, samples of mutations that yield viable offspring and populations in the long run, samples of habitats impacted by the molten lava of volcanoes, and samples in increases in growth in an economy and gradual changes that can yield large effects.[20] Here again, von Mises's view depends on the application of considerable knowledge before the statistics can be used.

This law depends then on ruling out from consideration samples having variations that cause a systematic change in a definite direction.

In the vast amount of data that is currently mined, only a very small portion appears to be suitable for use in a von Mises "collective" and that may more or less fit into the above strong law of large numbers. Since the computer has never produced an infinite number of results, so too as Keynes, Hume, and others have pointed out there has never been an infinite long run to test this strong law of large numbers.

---

[18] From Von Mises, 1957, Ibid., pp. 109–113, 116, 134. See pp. 163 and 199 in Carnap, Rudolf, 1962, *Logical Foundations of Probability*, Chicago: University of Chicago Press. It turns out that Carnap would agree: one can safely speak about the estimate and then and only then the ultimate outcome relative to the estimate.

[19] From Von Mises, 1957, ibid., pp. 80, 104, 105, and 108.

[20] Von Mises recognizes how small changes may accumulate to produce large effects, as when, in chaos theory, small changes in initial conditions have dramatic effects. 1957, Ibid., p. 180 and 182.

3.4 Weakening the Law(s) of Large Numbers

Recognizing that this strong law of large numbers is too strong, Von Mises further restates the weakened version of the strong law of large numbers:

> "if the game of no casts is repeated, and n is sufficiently large, *nearly all games* will yield the same value of the ratio n_1/n," the initial estimate of a relative frequency.

The expression "nearly all games" tells us that this is a matter of experience and that such long-run frequencies hold most of the time. The expression "nearly all games" is reminiscent of Kolmogorov's principles of inference (discussed in Chap. 2) in which he speaks about practical certainties. In terms of casting a die, von Mises himself details the manufacturing, toss techniques, and wear and tear that would significantly create "biases" in the die and hence the ultimate final result. Against the deductivists, von Mises can maintain that the die cast in these theoretical games is one that is not affected by such factors. Against his own views, von Mises is implicitly admitting that not even the casting of a die would in the long run produce a sequence of independent results. Like the deductivists, von Mises is postulating a "*p*" that has never been experienced.[21]

Von Mises also states a second weakened strong law of large numbers:

> If an object picked at random has shown a frequency of success a, in a long sequence of experiments, then the probability P that the probability p of this object lies between a-ε and a+ε will approach unity more and more closely as the number n of experiments is more and more increased.[22]

Restated, this second law is that "if the ratio $n_1/n=a$, and n is sufficiently large, nearly all evaluations must have approximately [the estimate] a." Here again, the expression "nearly all evaluations" indicates that von Mises is basing this result on experience.

These weakenings illustrate that Von Mises is having difficulties finding examples that fit the strong law of large numbers, samples that are equiprobable, random, independent, not trendy, not cyclic, not systemic. Optimists about data mining can be relieved because von Mises too finds little actual probabilistic purity—equiprobability, independence, randomness—in actual data.

The real world is full of various perturbations, shocks, trends, cycles, systemic relations, and data that relate to these phenomena. Common shocks to the system include stokes, dementia, malignancies, job losses, pension fund losses, huge unexpected financial losses, divorces, family deaths, rescues, auto accidents, lawsuits, and so on. Wars, assassinations, and economic downturns as well as home-team championships, major discoveries, and many other events may be epoch moments in one's life. Shocks and major epochs can yield major perturbations in one's lifestyle and modify what one has taken for granted through one's previous experience. Major changes may arise in one's expenditures, career goals, daily activities, cul-

---

[21] From Von Mises, 1957, Ibid., p. 127.

[22] From Von Mises, 1957, Ibid., pp. 122–125. Figure 1.1 in Chap. 1 illustrates how the confidence intervals converge to the estimate in question.

tural experiences, friends, and psychological well-being. What one may have taken for granted (e.g., severe winters) may be modified (seventy degrees Fahrenheit in the winter). One's previously narrow experience may in some such cases be greatly expanded.

3.5 A Major Object Lesson: Conditioning of Inferences that Deploy Large-Scale Data Management

The vast array of considerations in what data to use and how to use them do draw inferences is apparent in von Mises's serious concern about whether or not data are suitable for their use in probability and statistics. The law(s) of large numbers depend for their application on consideration of or presuppositions arising from many factors. The conditionality of inferences *from* data arises in part because there are so many considerations that enter into data management. These can pertain to the consideration of factors that can influence results, perturbations that may yield outliers, trends that underlie data, limits that proscribe extrapolations, and interferences in activities underway—let alone all the possible defects in errors themselves.

This conditionality is significant because inference is already going on before inferences are made from data. To illustrate this matter, an example from actuarial statistics provides some help. This conditional nature of statistics can also be shown if one attempts to take an extremely huge database, one collected over centuries and diverse circumstances, and not properly distinguish among these circumstances. For instance, one may take life-expectancy data from London in the 1600s and compare this with 1993 US life-expectancy data. Their combination would result in multimodality as in Fig. 3.2.[23]

This figure illustrates what happens when one first compares life-expectancy information from 1993 in the USA with the 1600s life-expectancy tables from London (derived from Bernstein 1996). These health environment contexts are so radically different that one would not expect to combine the two data sets without leading to something other than a bell-shaped curve. Context such as health environment used is critical in applying and understanding statistical data.

Thus, we can mine an immense amount of life-expectancy data, and if we put it all together, it is not likely to yield a normal distribution, which is long tailed, light tailed, and unimodal. Even today the inclusion of life-expectancy data from different regions will yield multimodality. There are many possible "perturbations" on these data.

If therefore one uses life-expectancy data, one needs to condition this data with a pertinent health environment and possibly with data pertaining to a personal genetics and history. The use of vast amounts of life-expectancy data before 1900

---

[23]From p. 83 in Bernstein, Peter L., 1996, *Against the Gods: The Remarkable Story of Risk*, New York: John Wiley & Sons, Inc.

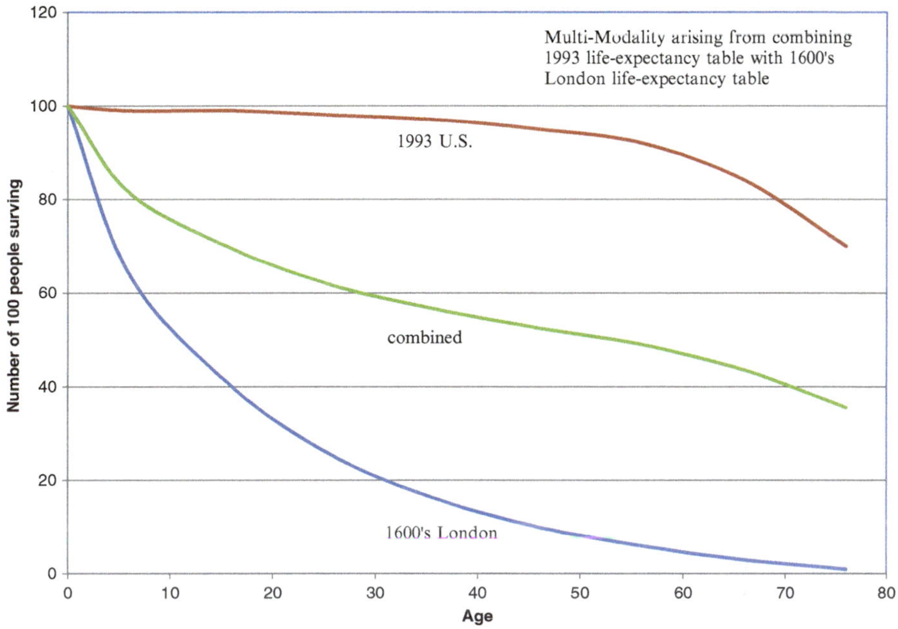

Fig. 3.2 Number of 100 people surviving (Y-axis) as a function of age and for two very different health environments 1993 USA

may yield very incorrect actuarial estimates if they had been used to forecast the twentieth century and the twenty-first century actuarial rates. The use of conditioning on statistics yields potentially diverse perspectives. For instance, if one is looking at the life expectancy of a female at age 50, one may use mass statistics in general. Alternatively, one may choose statistics for the particular person given a genetic background and personal health and environmental history. Given that there are different ways of managing the data on a person in order to arrive at a conclusion, as J. M. Keynes asks, since "a given proposition belongs to innumerable different classes, how are we to know which class the premises indicate as appropriate?"[24]

3.6 Summary of the Frequency Theory

Von Mises may accept most of the following conclusions from a review of his account of frequency theory. In particular, he is keenly aware of how many applications of statistics may not be as cautious as desirable. Nonetheless, the lessons learned from an evaluation of his theory are many and not all corroborate his approach nor the optimism of some data-mining activities:

[24] From Keynes, 1921, op. cit., p. 103.

1. Curation and large extant data sets illustrate how results derived are conditional on management and selection of pertinent data. Considerable time and effort—inference—is required in these management processes. These may not result in unambiguous results.

   This first conclusion arises from the consideration of data used to diagnose a person's health given that person's health history and the vast amount of data on health. This vast information depends for its application in this case on relevant populations, not necessarily entire populations, and the person's genetics and history too may contain data that are misleading and that need to be interpreted provisionally and holistically. The vast amount of data that can in principle be available is valuable for application under these conditions.

2. Simulation procedures can employ say laws of large numbers owing to the ability to provide quasi-random numbers and, with sometimes considerable effort, to develop equiprobable estimates based in model inputs (that may themselves have variable probabilities). Under many conditions (discussed especially in Chap. 5), one can employ laws to derive conclusions for which the law(s) of large numbers obtain.

   These simulation procedures can be enormously helpful in developing conclusions from data, models, and assumptions derived from considerable effort. Using frequency theory, one can derive confidence intervals. Using enormous numbers of simulations, one can avoid undue simplifications that have been employed before vast IT capabilities have been present.

   However, as von Mises's quest for pure collectives suggests, real-world equiprobable cases that are independent of others in the collective and hence are not part of cycles, trends (radioactive elements), systems, and the like are rarer than one may first suppose. For real-world cases, there is a strong urge to *assume* stabilities, whether in most or all cases. To assume stabilities is to assume that the future will be like the past. Very much concerned to find stabilities, J. M. Keynes maintains that stable frequencies are not very common and cannot be assumed lightly.[25] Hundreds of millions of data on life expectancy in the past do not always indicate the nature of trends for the population as a whole or for subgroups within the population. In many of these cases, what has been taken for granted may need to be reevaluated, sometimes extensively. Just as past individual experience—narrow as it typically is—may become expanded, so expectations about growth, priorities, and many other matters may become modified as a result of shocks.[26]

3. Mega-risks can occur to infrastructure systems either regarded in isolation from the rest of the world. Low rates of pipeline replacement may over time eventually yield clusters of trunk line ruptures in a culinary water system. Previous experience with needed pipeline repairs may be upset such an outbreak of ruptures. In

[25] From Keynes, 1921, op. cit., p. 335.

[26] In 1962, on p.188 of *The Silent Spring*, Boston: Houghton Mifflin Company, Rachel Carson maintained that "it is simply impossible to predict the effects of lifetime exposure to chemical and physical agents that are part of the biological experience of man."

electric power systems, the additions of many portable and laptop computers may significantly increase the number of watts needed for a specific population and outpace what had estimated in long-term planning. Outages and brownouts may result from major increases in demand. New health cures may likewise change healthcare systems dramatically. Life expectancies before the twentieth century—even with the result of very long sequences—may be modified significantly as new drugs are devised. Still, with population increases, famines, plagues, insecticides, and other phenomena may greatly reduce life expectancies. Catastrophes, of course, can disrupt infrastructure systems for many years.

4. As J. M. Keynes (also David Hume, Francis Bacon, Gottfried Leibniz, among others) have proclaimed, there have been no cases in which an infinite number of samples have confirmed empirical induction. Empirical induction is first of all not mathematical induction. Reference to previous successes belies the dependence on either long-term mathematical convergence of results or a finite number of cases use them to draw conclusions that have turned out to be adequate. Frequency theory requires in general that finite samples are used to develop conclusions. (As already stated, a very large number of conclusions, assumptions, and the like are required before one can even define the finite samples to be used in drawing inferences.)
5. The frequency theory can thus provide enormous assistance in addressing the original questions raised in these chapters. It is conceded throughout this chapter that many loss distributions are light tailed. However, its serious limitations suggest that it is not to be taken as the entire answer. The version of this theory provided by von Mises provides enormous insights into the difficulties of using laws of large numbers for the vast majority of data-mining activities, especially those pertaining to shocks to systems (such as the long-term shocks revealed by Rachel Carson and many others).

The quest for a set of purely random, equally likely, independent samples needed for data mining to ensure valid statistical conclusions reminds one of mythical attempts to fly on Pegasus to Mt. Olympus, only to suffer serious downturns in this quest.[27] Von Mises's insights in his view of frequency theory helps to deflate what statistics can do, somewhat excessively, but even in data mining the absence of infinitely many samples and the errors that can arise with data not yet gathered entail that the mythical top of Mt. Olympus is unattainable. Both frequency theory and data mining, though, greatly enhance our knowledge through the quest itself.

[27] As a result of all the practical activities that those engaged in data mining participate in, they do not need to worry about having the fate of Bellerophon, who, after having been a hero who killed the Chimera, tried to fly to Mt. Olympus on Pegasus, was thrown off Pegasus by Zeus, and "wandered alone about the plain of Aleios, eating his heart out, skulking aside from the trodden track of humanity." (From The Iliad, book VI, translated by Richard Lattimore, Chicago: Phoenix Books, University of Chicago Press, 1951).

References

Anderson, C. (2008). The end of theory: The data deluge makes the scientific method obsolete. *Wired,* 6/23/08.

Bell, G. (2009). Foreword. In T. Hey, S. Tansley, & K. Tolle (Eds.), *The fourth paradigm: Data-intensive scientific discovery* (pp. xiii–xvii). Redmond: Microsoft Research.

Bernstein, P. L. (1996). *Against the Gods: The remarkable story of risk.* New York: Wiley.

Cambel, A. B. (1993). *Applied chaos theory: A paradigm for complexity.* San Diego: Academic. A. B Cambel states.

Carnap, R. (1962). *Logical foundations of probability.* Chicago: University of Chicago Press.

Carson, R. (1962). *The silent spring.* Boston: Houghton Mifflin Company.

Diamond, J. (2005). *Collapse: How societies choose to fail or succeed.* New York: Viking.

Efron, B., & Tibshirani, R. J. (1993). *An introduction to the bootstrap.* New York: Chapman & Hall.

Ford, J. (1983). How random is a coin toss? *Physics Today, 36*, 40–47.

Gannon, D., & Reed, D. (2009). Parallelism and the cloud. In T. Hey, S. Tansley & K. Tolle (eds.), *The fourth paradigm: Data intensive scientific discovery.* Redmond: Microsoft Research.

Hey, T., Tansley, S., & Tolle, K. (Eds.). (2009). *The fourth paradigm: Data-intensive scientific discovery.* Redmond: Microsoft Research.

Keynes, J. M. (1921). *Treatise on probability.* London: MacMillan & Co.

Lattimore, R. (1951). *The Iliad of homer in a new verse translation.* Chicago: University of Chicago Press.

Van De Sompel, H., & Lagoze, C. (2009). *All aboard: Toward a machine-friendly scholarly communication system.* Redmond: Microsoft Research.

von Mises, R. (1957). *Probability, statistics and truth.* New York: Dover Publications.

Von Plato, J. (1994). *Creating modern probability: Its mathematics, physics and philosophy in historical perspective.* Cambridge, MA: Cambridge University Press.

Chapter 4
Subjectivist (Bayesian) Theory

...something like "convergent thinking" is just as essential to scientific advance as is divergent. ...most new discoveries and theories in the sciences are not merely additions to the existing stockpile of scientific knowledge. To assimilate them the scientist must usually rearrange the intellectual and manipulative equipment he has previously relied upon, discarding some elements of his prior belief and practice while finding new significance in and new relationships between many others....the ultimate effect of this tradition-bound work [puzzle solving] has been to change the tradition. (From pp. 226, 227, 234 in "The Essential Tension: Tradition and Innovation in Scientific Research," pp. 225–239 in Kuhn, Thomas, 1977, The Essential Tension: Selected Studies in Scientific Tradition and Change, *Chicago: University of Chicago Press)*

Tetlock interviewed 284 people who made their living "commenting or offering advice on political and economic trends." The Results were devastating. The experts performed worse than they would have if they had simply assigned equal probabilities to each of the [three] alternatives. ... Even in the region they knew best, experts were not significantly better than non-specialists. (From p. 219 in Kahneman, Daniel, 2011, Thinking, Fast and Slow, *New York: Farrar, Straus and Giroux and based on Tetlock, P. E., 2005,* Expert political judgment, *Princeton, N. J.: Princeton University Press)*

Abstract This chapter presents the third major view of probability and statistics, the subjectivist or Bayesian theory. The prior two chapters emphasized a linear theory proceeding from known to unknown. In contrast, the Bayesians emphasize updating beliefs, even beliefs that may not be "known," but which provide a starting point in solving problems using probability and statistics. Like the frequentist theory, the Bayesian theory heavily depends on the use of data, or facts, that permit the Bayesians to contend that they have used facts in order to obtain adequate solutions to problems. Because both frequency theory and Bayesian theory have perceived successes, they vie with each other in present-day popularity as the two and only two viable views of probability and statistics. However, as with the frequency theory, the Bayesian method is inadequate for the purposes of this book because it assumes a convergence at infinity which will never be experienced.

C.E. Taylor, *Robust Simulation for Mega-Risks*, DOI 10.1007/978-3-319-19413-4_4

4.1 The Bayesian Theory Contrasted to Two Other Major Probability Theories

This chapter presents the third major view of probability and statistics, the Bayesian theory, occasionally called subjective. The chief sense in which the Bayesian theory is "subjective" is that it is nonlinear. The Bayesian theory heavily depends on the use of data or facts, which permit the Bayesian to contend that they have used these facts in order to obtain adequate solutions to problems. In contrast to the theories in the prior chapters, Bayesians emphasize updating beliefs, and so are happy to begin with beliefs that may not be "known," but provide a starting point in solving problems using probability and statistics.

The Bayesian view, as this chapter will indicate, thrives on successes. The book *The Theory that Wouldn't Die* by the journalist Sharon McGrayne, for instance, provides an enormous number of purported successes. Often these involve secret missions that need to solve cases in political conflicts, such as how to locate U-boats in war. However, the main thrust in McGrayne's work appears to cover cases—whether secret or public—in which searches are greatly facilitated through the use of Bayesian statistics in contrast to the more tedious cumulative approaches in frequency theory.[1] Because both frequency theory and Bayesian theory have claimed successes (also noted in Chap. 6 for work using Fisher and/or Pearson), they vie with each other currently in popular terms as the two and only two viable views of probability and statistics.

4.2 Overview of the Third Main Theory of Probability

Bayesians do not hold that probability statements reflect relative frequencies. Instead, for Bayesians, probability statements express degrees of belief. Randomness and probability exist only in the minds of individuals, not in the objective world. Probabilities thus express the "*degree of belief* in the occurrence of an event attributed by a given person at a given instant and with a given set of information." P(H|E) is regarded as a measure of my degree of belief in H "given that I know that E is true."[2]

[1] See McGrayne, Sharon Bertsch, 2011, *the theory that would not die: how Bayes' rule cracked the enigma code, hunted down Russian submarines & emerged triumphant from two centuries of controversy*, New Haven: Yale University Press.

[2] On David Hume's version of "belief," see Bennett, Deborah J., 1998, *Randomness*, Cambridge, Massachusetts: Harvard University Press, p. 154; for Bayesian accounts of the quotation, see De Finetti, Bruno, 1970, *Theory of Probability: A critical introductory treatment*, Chichester West Sussex: John Wiley & Sons, Wiley Classics Library Edition published 1990, pp. 3,4; De Finetti, Bruno, 1937, "Foresight: Its Logical Laws, Its subjective Sources," republished pp. 53–118 in *Studies in Subjective Probability*, edited by Henry E. Kyburg, Jr. and Howard E. Smokler, Huntington, New York: Robert E. Krieger Publishing Company, 1980, p. 109; Von Plato, J., 1994,

The Bayesian view is that a subject's prior judgment, even though fallible and not even necessarily "known," can enhance the extant data so that issues can be addressed more expeditiously than with frequency theory procedures. This view represents a nonlinear view of the development of statistical findings. To repeat, the Bayesian recognizes that the subject's prior view is fallible and extremely corrigible, but this is not an overwhelming reason not to use the probabilistic value that it contains. For the Bayesian, knowledge progresses when a corrigible prior judgment is combined with an incomplete database. Adequate solutions to probability and statistical problems can arise through this nonlinear Bayesian approach. As with the data-mining activities that von Mises requires in order to set up collectives, the Bayesian theory requires a give-and-take movement between hypotheses and empirical samples.

Against this background, this chapter first provides a discussion of the derivation of Bayes theorem from the theorem of total probability. The latter is expressed in terms of a *finite* partition of a universal set. Next are some simplifications from Nate Silver on how to calculate Bayesian estimates, along with reference to some fairly simple illustration that Silver employs.

Since the heavy emphasis of these essays is on developing mega-risk distributions, there is a discussion of how one might use Bayesian procedures to estimate the occurrence of a catastrophic peril, in this case major earthquake occurrences. This discussion leads into an account of the challenge of how different people can have different degrees of belief and how this is alleged to be overcome through a theory of large samples. Finally, there is a summary of pros and cons of Bayesian theory.

It has clearly had many valuable applications. For instance, one might have a prior distribution for the roulette wheel: if one picks black, one has a 36/73 chance of winning. If there is some physical anomaly in the roulette wheel, this should surface in the many applications of this table relative to the prior expectations. Many games of chance, for instance, can be viewed in similar terms. Sharon McGrayne's work provides an abundance of illustrations.[3]

Yet the categorical version of Bayesian theory encounters the same problems as does the frequency theory. Only in later chapters do we discern how the Bayesian theory, as well as the other two major theories, can assist in addressing the initial question with which this inquiry begins.

Creating Modern Probability: Its Mathematics, Physics and Philosophy in Historical Perspective, Cambridge: Cambridge University Press, p. 24; Lee, P. M., 2004, *Bayesian Statistics: An Introduction*, New York: Oxford University Press Inc., Third Edition, p. 3; see also Press, S. J., 1989, *Bayesian Statistics*, New York: John Wiley & Sons, pp. 3, 4.

One of the major issues with this characterizations of probability and statistics is that it confines itself to an "individual's" set of beliefs, and as this chapter continues, one sees that the alleged convergence between an individual's set of beliefs, taken initially in isolation, and that of others requires postulating a long run that in Keynes's words never arrives.

[3] The illustration of the roulette wheel comes from Mlodinow, Leonard, *The Drunkard's Walk: How Randomness Rules Our Lives,* New York: Pantheon Books, an e-book. For many more illustrations, see again McGrayne, Sharon Bertsch, 2011, op. cit.

4.3 The Derivation

The finite perspective of Bayesian modeling begins with the definition of a "partition" below[4]:

A **partition** of U is a division into mutually exclusive sets, A_i, such that the following three conditions obtain.

For all integers $i, j > 0$:

1. $P(A_i \cap A_j) = \phi$.
2. $\cup_{j=1}^{n} (A_j) = U$
3. This definition requires that $P(A|B_i) > 0$.

Hence, the partitions used in Bayes Theorem cannot contain events of probability 0. This can happen for a continuous distribution if some B is a point. However, there are probability measures for which points have probability greater than zero.[5]

The **theorem of total probability** thus asserts the following:

Assume that $\{B_j\}$ constitutes a partition of U, then

$$P(A) = (A \mid B_1) P(B_1) + P(A \mid B_2) P(B_2) + \cdots + P(A \mid B_n) P(B_n) \tag{4.1}$$

According to Meyer,[6] in order to calculate P (A), one needs to know B_j for each j.

The definition of ***conditional probability*** *is as follows:*

For any two events X and Y, $P(X \mid Y) = P(X \cap Y)/P(Y)$
and $P(Y \mid X) = P(Y \cap X)/P(X)$
Thus, $P(X \mid Y) * P(Y) = P(Y \mid X) * P(X)$

From this definition of conditional probability and the theorem of total probability, one derives **Bayes' theorem**:

$$\begin{aligned} P(B_i \mid A) &= P(A \mid B_i) * P(B_i) / \left[\sum P(A \mid B_j) P(B_j) \right] \\ &= P(A \mid B_i) * P(B_i) / P(A) \end{aligned} \tag{4.2}$$

In this formulation, the **prior probability** is $P(B_i)$ and the **likelihood function** is $[P(A|B_i)]$. The **posterior probability** is $P(B_i|A)$. This definition only has value if $[P(A|B_i)] > 0$. Hence, this definition requires that one does not "partition" possible

[4] The illustration of the roulette wheel comes from Mlodinow, Leonard, *The Drunkard's Walk: How Randomness Rules Our Lives*, New York: Pantheon Books, an e-book. For many more illustrations, see again McGrayne, Sharon Bertsch, 2011, op. cit.

[5] Dr. Robert Riehemann, Letter of December 22, 2014.

[6] This treatment is from Meyer (1970).

events in such a way (as through using "points") that yield probabilities that are virtually zero.

If P(A) is treated as a constant, then one has the following very useful formulation:

$$P(B_i \mid A) \propto P(A \mid B_i) * P(B_i) \tag{4.3}$$

Furthermore, the theorem does not permit an indefinitely large partition of the universal set. Thus, the partition does not envisage application to a large number of new prospective trials or experiments. If the experiment to be predicted is to be incorporated into Bayes' theorem, then the theorem of total probability contains an event or events (here, B_i, and by implication, A) whose probabilities are unknown.

When the prospective events to be updated, say m, exceed the original "n" samples (or n−1, if the nth sample is to be estimated), then each future update must have an additional event included. This potential infinite progression illustrates how the original formulation of Bayes' theorem demands a finite point of view.[7]

4.4 Simplified Version and Illustrations by Nate Silver

The esteemed bettor Nate Silver has developed the following simplifications:

$$P(B) = x$$
$$P(A \mid B) = y$$
$$P(A \mid \text{not} - B) = z$$

from which he derives

$$P(B \mid A) = xy / (xy + z(1 - x)) \tag{4.4}$$

If one ignores subscripts, Eq. (4.4) follows from Eq. (4.2) and yields Eq. (4.3). Using this simplification, Silver discusses the following sorts of cases:

- The estimation that a women is cheating based on finding her underwear
- The estimation that women at age 40 should have mammograms

[7] There are of course continuous models that might be taken to account for denumerably many or even non-denumerably many samples. For instance, an integrand formulation of Bayes' estimator is presented on p. 349, Hoel, Paul G. 1971, *Introduction to Mathematical Statistics*, New York: John Wiley & Sons, Inc. On reflection of the notion that temperatures, for instance, are continuous, one may still be stuck with (a) point estimates or (b) line segments for temperature. In general, the number of samples from a line or area or volume will be finite, even though under some circumstances it may even be "necessary" to consider irrational numbers in estimating their lengths, areas, or volumes, respectively.

- The estimate of the probability of terror attack given a first plane hitting the World Trade Center
- The estimate of the probability of terror attack given a second plane hitting the World Trade Center

For Silver, the latter two examples turn out to have probabilities of 38 % and 99.99 %, respectively. The second example serves to illustrate why mammograms for women under 50 have been controversial.[8] These examples show some of the wide range of examples for which Bayesian estimates can be made.

4.5 Illustration of Estimating the Distribution of Occurrence of a Rare Catastrophe Peril

A detailed example of the use of a Bayesian approach comes from D. Perkins,[9] who claims that one can estimate probabilities of occurrence of some infrequent peril (here, major earthquake occurrences) as follows:

Use, for example, a Poisson distribution as one's likelihood function. Like the binomial model, a Poisson distribution has no "statistical memory" and so is unlike, for instance, a disease that may depend on the eventual accumulation of harmful doses.

The Poisson probability for number of events in time, t, is expressed as follows[10]:

$$P(E = J) = (\lambda t)^J \frac{e^{-\lambda t}}{J!} \tag{4.5}$$

in which

E = a variable for the number of events in time interval t
λ = the mean rate of occurrence for each time unit
$J = 0,1,2,\ldots$

Both the mean and the variance equal λ.

---

[8] These illustrations are found in Nate Silver, 2012, *The Signal and the Noise,* New York: the Penguin Press, pp. 242ff. Textbooks on Bayesian statistics can provide a large number of other valuable illustrations. The mammogram examples appear as well on pp. 259–261 in McGrayne, Sharon, 2011, op. cit.

[9] This illustration comes from David Perkins, written comm.., July 2010.

[10] See, for instance, Ang, Alfredo H-S. and Wilson H. Tang, 1975, *Probability Concepts in Engineering Planning and Design, Volume I, Basic Principles,* New York: John Wiley & Sons, pp. 114ff. and Law, A. M. and Kelton, W. D., 1991, *Simulation Modeling and Analysis*, New York: McGraw-Hill, pp. 349ff.

For the example in question, the Poisson model is used for the likelihood function, $P(A|B_i)$, expressed as follows for three events in 1200 years:

$$Y = (2/1200) * (x * 1200)^3 * \exp(-x * 1200), \quad \text{in which } x \text{ is the rate of occurrence.} \tag{4.6}$$

For the geological prior $P(B_i)$, this rate is expressed as follows:

$$Y = \frac{(8.3/10^6)}{\sqrt{2\pi\sigma^2}} \frac{1}{x} e^{-\frac{1}{2}\left(\frac{\log(x)-\log(0.004)}{0.4}\right)} \tag{4.7}$$

4.6 is the likelihood for rate x given three events in 1200 years. 1200x is the expected number of events, which is the parameter for the Poisson distribution. 4.7 is the prior, which is a lognormal distribution whose location parameter is 0.004, which is the equivalent of 4 in 1000.

The parenthetical values leading each equation have no function other than to give the same peak height in the graphic calculator program I use. So, the 2 and the 8.3 are just adjustable parameters.[11]

Table 4.1 assists in clarifying how one simulates the development of the posterior from these two equations. In the first column, one begins with $x=0.0005$, then proceeds to $x=0.0005$ times $2=0.001$, and so on. The column ends at 0.0135 since beyond this point, only very small numbers result in other columns.

The second column uses Eq. (4.7) for the geological prior. The estimates are "raw" in the sense that their total is not equal to one. The third column uses Eq. (4.6) for the likelihood function. Again, these are "raw" estimates based directly on the equation above.

Next one normalizes these results for the Y's as calculated so that they sum to 1.0. Columns four and five cover normalized estimates of the prior and likelihood function, respectively.

Column six multiplies columns four and five together as a version of Eq. (4.2) to obtain one set of "raw" estimates of the "posterior" function. Column seven produces a normalized version of this estimate of the posterior function. Note that there are different ways to weight the likelihood function and the prior function. For instance, Gelman and others discuss using variances to "weight" the prior and likelihood functions. The simple illustration here weights these two functions equally.[12]

[11] David Perkins email of May 12, 2014.

[12] Gelman, Andrew, John B. Carlin, Hal S. Stern, and Donald B. Rubin, 2003, *Bayesian Data Analysis*, Boca Raton: Chapman & Hall/CRC., p. 47.

Table 4.1 Calculations, for example, of Bayesian analysis (very small values omitted can have cumulative effects)

X	Raw prior total	Raw likelihood total	Prior normalized	Likelihood normalized	Raw posterior	Normalized posterior
0.0005	0.224312	0.000198	0.226752	0.011852	0.002688	0.324763
0.001	0.049847	0.000867	0.050389	0.052038	0.002622	0.316860
0.0015	0.016635	0.001607	0.016816	0.096386	0.001621	0.195861
0.002	0.006231	0.00209	0.006299	0.125388	0.00079	0.095436
0.0025	0.002292	0.00224	0.002317	0.134403	0.000311	0.037627
0.003	0.000716	0.002125	0.000723	0.127461	0.000092	0.011141
0.0035	0.000132	0.001852	0.000134	0.111081	0.000015	0.001793
0.004	Very small	0.001517	Very small	0.090999	Very small	Very small
0.0045	"	0.001185	Very small	0.071108	"	"
0.005	"	0.000892	"	0.053532	"	"
0.0055	"	0.000652	"	0.039104	"	"
0.006	"	0.000464	"	0.027862	"	"
0.0065	"	0.000324	"	0.019441	"	"
0.007	"	0.000222	"	0.013326	"	"
0.0075	"	0.00015	"	0.008995	"	"
0.008	"	Very small	"	0.005991	"	"
0.0085	"	"	"	0.003944	"	"
0.009	"	"	"	0.002569	"	"
0.0095	"	"	"	0.001658	"	"
0.01	"	"	"	0.001062	"	"
0.0105	"	"	"	0.000674	"	"
0.011	"	"	"	0.000426	"	"
0.0115	"	"	"	0.000267	"	"
0.012	"	"	"	0.000166	"	"
0.0125	"	"	"	0.000103	"	"
0.013	"	"	"	Very small	"	"
0.0135	"	"	"	"	"	"

Figure 4.1 summarizes the procedure. One notices that the posterior function begins with the highest values, and the likelihood function has higher values as x increases. The posterior function is the probability density distribution of the multiplication of the probability density function (pdf, with values summing to one) for the prior function and the pdf for the likelihood function. Note as stated already that the prior pdf and likelihood pdf are weighted equally in the figure.

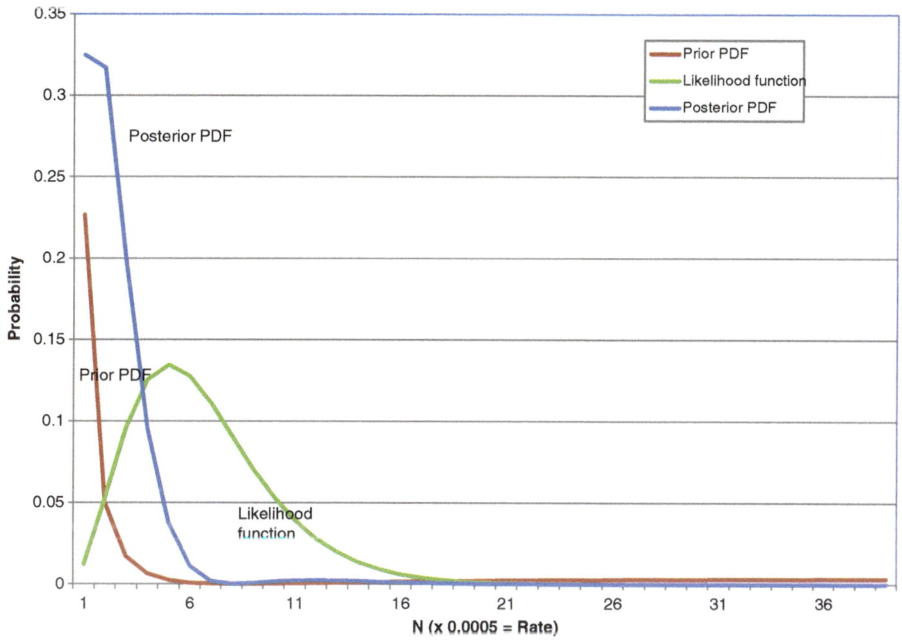

Fig. 4.1 Estimate probabilities of occurrence of some infrequent peril (here earthquakes) (Source: David Perkins)

4.6 Slightly Divergent Prior Estimates: Multiple Models

The first quotation in this chapter stresses views by T. S. Kuhn on how the educational process begins with an emphasis on "convergence," with "divergence" later in such sciences as physics. For instance, one learns that Kepler's law and, with revised data by astronomers, Newton's inertial rectilinear motion plus gravitation law provide an excellent first approximation to show that planets have elliptical orbits about the sun. This is elaborated in Chap. 9.[13]

The second quotation by P. Tetlock stresses how, in the difference between hedgehogs (one big view) and foxes (many small views), so-called political experts do not fare well. (Note that hedgehogs appear to be drawn from neocons and like-minded people.) Hedgehogs on this view tend to be inflexible yet divergent from foxes.

[13] The initial account can be found in Feynman, Leighton, Sands, 1963–2010, pp. 7–2, 7–3 to be expanded in Chap. 9 on this book.

Every day, there are divergent opinions, viewpoints, calibrations, and the like that can yield different prior distributions. For instance, one person analyzing the 2012 Presidential Election may bet that Obama will obtain 290 Electoral College votes, another may bet that he will obtain 236 electoral collect votes, and a third may estimate that he will obtain 330 Electoral College votes (the number that Nate Silver bet on and that turned out to be accurate).

For another example, one person may maintain that mature faults have fixed boundaries or termini and have a return interval that is fixed, with statistical uncertainty. Another person may maintain that those boundaries for mature faults may be violated when an earthquake ruptures first along the primary fault and then diverges along a secondary fault. There are a great many such divergent opinions in earthquake estimation procedures.[14]

For the sake of illustration, the next figure outlines how slightly divergent estimates for the prior distribution can yield multiple posterior distributions. In particular, the original prior Eq. (4.7) is:

$$Y = (8.3/1000000) * (1/x) * \exp\left(-0.5 * \left(\left(\ln(x)\ \ \ln(0.004)\right)/0.40\right)^2\right)$$

One slightly divergent equation is:

$$Y = (8.3/1000000) * (1/x) * \exp\left(-0.5 * \left(\left(\ln(x)\ \ \ln(0.003)\right)/0.40\right)^2\right) \quad (4.8)$$

A second slightly divergent equation is:

$$Y = (8.3/1000000) * (1/x) * \exp\left(-0.5 * \left(\left(\ln(x)\ \ln(0.005)\right)/0.40\right)^2\right) \quad (4.9)$$

As expected, Fig. 4.2 shows that the resulting posterior distribution is different for the three slightly divergent prior models. This obvious conclusion shows how diverse individuals with slightly different prior models can arrive at divergent outcomes or posterior distributions.

Of special note is the more often contested contention that similar results obtain if the likelihood function is different for different individuals. This indeed has often been the case, as "data" are themselves often contested, or found to be irrelevant, for evaluations at hand.

[14] For a discussion of divergences for California seismicity, see Lee, Yajie, Craig E. Taylor, Zhenghui Hu, William P. Graf, Charles K. Huyck, 2014, "Uncertainty estimates for earthquake hazard analysis through Robust Simulation," National Earthquake Research Conference, hosted in Anchorage, AK, by the Earthquake Research Engineering Institute.

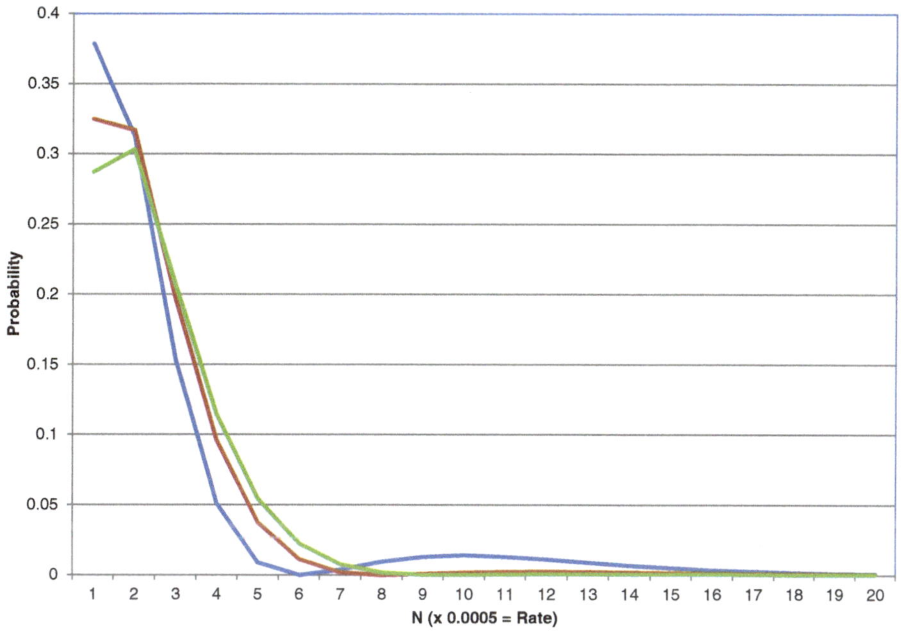

Fig. 4.2 Comparison of three posterior distributions (first 20 top values of 250 possible values) with slightly divergent prior models

4.7 Convergence in the Long Run

For the most part, the divergences that one finds from smaller samples is thought to vanish once larger samples are found. The likelihood function becomes dominant. That is, the frequentist position and the Bayesian position converge (see Fig. 1.1 in Chap. 1 for this gradual convergence). According to Lee,

> It is often sensible to analyze scientific data on the assumption that the likelihood dominates the prior. ...even if you and I both have strong prior beliefs about the value of some unknown quantity, we might not agree, and it seems sensible to use a neutral *reference prior* which is dominated by the likelihood and could be said to represent the views of someone who (unlike ourselves) had no strong beliefs a priori. The difficulties of public discourse in a world where different individual have different prior beliefs constitute one reason why a few people have argued that, in the absence of agreed prior information, we should simply quote the likelihood function...in many scientific contexts we would not bother to carry out an experiment unless we thought it was going to increase our knowledge significantly, and that in the case then presumably the likelihood will dominate the prior.[15]

[15] See Lee, P. M., 2004, *Bayesian Statistics: An Introduction*, New York: Oxford University Press Inc., Third Edition, p. 43.

Gelman and others concur that except in some rare cases, there is convergence in the long run.[16]

If sound, these views overcome, for instance, the extreme views, as expressed by de Finetti, that probability is "the opinion of an individual and cannot have meaning except in relation to him."[17] If the likelihood function dominates, then the probability ceases to be dominated by the opinion of any individual.

4.8 Summary of Subjectivist (Bayesian) Theory

4.8.1 Upsides

As with the frequency theory, the Bayesian theory has had many applications and has proven to be attractive to many practitioners. Many tools have developed to simplify these applications. In principle, all the tools developed for the frequency approach should be adaptable to the Bayesian approach, especially as regards the application of the likelihood function.

Of special interest especially to later chapters is how individual people can contribute to statistical estimates over and above extant data. The implication is that even individual people's views, treated here as "priors," can be cognitively valuable even though they may not be entirely true. At first blush, the Bayesians do not demand a wholesale elimination of a priori assumptions relative to specific inquiries. Even though the accumulation of data contribute to updates provided within Bayesian theory, the theory does not start from firm foundations and so is nonlinear. This nonlinear view of statistics and probability continues throughout these essays and benefits from the struggles of Bayesians to promote their views.

There is nonlinearity implicit in von Mises's search for collectives, presumably by data-mining activities along with tests. Many conclusions are discovered during this data-mining and curation activity. However, this nonlinearity is a consequence of his view rather than something that he discusses at length.

There are also indications that Bayesian models may expedite reaching solutions. Von Mises complains that Bayesian evaluations apply only to small samples. This may turn out to be a great advantage rather than a disadvantage. Sharon McGrayne, for instance, finds arguments that in some cases very small samples are "enough" using Bayesian modeling, whereas frequency modeling requires a great many more. In addition, Bayesian modeling is regarded as being valuable in the many instances in which there are no data yet to make a decision, as often occurs in business.[18]

---

[16] Gelman, Andrew, John B. Carlin, Hal S. Stern, and Donald B. Rubin, 2003, *Bayesian Data Analysis,* Boca Raton: Chapman & Hall/CRC., pp. 107–112.

[17] De Finetti, Bruno, 1937, "Foresight: Its Logical Laws, Its subjective Sources," republished pp. 53–118 in *Studies in Subjective Probability*, edited by Henry E. Kyburg, Jr. and Howard E. Smokler, Huntington, New York: Robert E. Krieger Publishing Company, 1980, p. 109.

[18] Only a small sample of pertinent passages from Sharon McGrayne, 2011, op. cit. include pp. 65, 85, 92, 94, 141.

4.8.2 Downsides

As with the frequency theory of probability and statistics, the resort to a presumed convergence in the long run implies that one has indeed had confirmations of frequencies in the very long run. Again, as J. M. Keynes has intimated following David Hume, the theory proposed actually has a null set of confirmations.[19] No one ever experiences all the trials of any infinite sequence. Thus, contending that there are tests for randomness or that past experience confirms that "random sequences" always yield a unique stable probability p has no evidence whatsoever.

Stresses on convergence among "rational" individuals as well as among ultimate outcomes reached by the application of frequency theory stresses how especially Chaps. 3 and 4 are dominated by the view that facts are facts and remain so forever. Even with "curated" facts, though, some may be contested. And other "facts" may be deemed to be irrelevant or weighted too heavily by specific investigative teams.

Even though Bayesian theory has nonlinear elements and brings in an active role by individuals, the claims that people will eventually agree in the long run presuppose that individuals will have experienced a long run. In science, engineering, and other heavily quantitative subject matters, agreement on all details, including all approaches, is not necessarily the rule—even in the long run.

Bayesian methods are treated as inverse modeling, an enormously valuable activity. However, as with other comments on convergence, inverse modeling likewise is discussed later in, for instance, Chap. 6. While most desirable for many problems, inverse modeling only rarely yields single or convergent solutions.

References

Ang, A. H.-S., & Tang, W. H. (1975). *Probability concepts in engineering planning and design* (Basic Principles, Vol. I). New York: Wiley.

Bennett, D. J. (1998). *Randomness*. Cambridge, MA: Harvard University Press.

De Finetti, B. (1937). Foresight: Its logical laws, its subjective sources. In H. E. Kyberg & Smokler, H. E. (eds.), *Studies in subjective probability*. New York: Robert E. Krieger Publishing Company.

De Finetti, B. (1970). *Theory of probability: A critical introductory treatment*. Chichester: Wiley, Wiley Classics Library Edition published 1990.

Gelman, A., Carlin, J. B., Stern, H. S., & Rubin, D. B. (2003). *Bayesian data analysis*. Boca Raton: Chapman & Hall/CRC.

Hoel, P. G. (1971). *Introduction to mathematical statistics*. New York: Wiley.

Kahneman, D. (2011). *Thinking, fast and slow*. New York: Farrar, Straus and Giroux.

Keynes, J. M. (1921). *A treatise on probability*. London: MacMillan and Co.

Kuhn, T. (1977). *The essential tension: Selected studies in scientific tradition and change*. Chicago: University of Chicago Press.

---

[19] From Hume, op. cit., pp. 49ff. and Keynes, John Maynard, 1921, *A Treatise on Probability*, London: MacMillan and Co, p. 82.

Lee, P. M. (2004). *Bayesian statistics: An introduction* (3rd ed.). New York: Oxford University Press.

Lee, Y. J., Taylor, C. E., Hu, Z., Graf, W. P., & Huyck, C. K. (2014). Uncertainty estimates for earthquake hazard analysis through robust simulation. In *Tenth U.S. National Conference on Earthquake Engineering Frontiers of Earthquake Engineering*, July 21–25, 2014 10NCEE Anchorage, Alaska.

McGrayne, S. B. (2011). *The theory that would not die: How Bayes' rule cracked the enigma code, hunted down Russian submarines & emerged triumphant from two centuries of controversy*. New Haven: Yale University Press.

Meyer, P. L. (1970). *Introductory probability and statistical applications*. Reading: Addison-Wesley Publishing Company.

Press, S. J. (1989). *Bayesian statistics*. New York: Wiley.

Silver, N. (2012). *The signal and the noise*. New York: The Penguin Press.

Tetlock, P. E. (2005). *Expert political judgment*. Princeton: Princeton University Press.

Von Plato, J. (1994). *Creating modern probability: Its mathematics, physics and philosophy in historical perspective*. Cambridge: Cambridge University Press.

Part III
The Dilemma of "Infinity"

Chapter 5
The "Wobble" Breaks Through Previous Theories (Mid-1970s)

Consistent statistics ...all tend more and more nearly to give the correct values, as the sample is more and more increased; at any rate, if they tend to any fixed value it is not to an incorrect one. (From Fisher, Ronald Aylbumer, 1944, Statistical methods for research workers, *London: Oliver and Boyd Ltd., ninth edition, p. 11)*

Mathematicians, however, are well aware that it is childish to try to show that every continuous function has a derivative. Though differentiable equations are the simplest and easiest to deal with, they are exceptional. (Quoted from Jean Perrin on p. 8 in Mandelbrot, Benoit B., 1983, The Fractal Geometry of Nature, *New York: W. H. Freeman and Company, originally 1977. On p. 178 in his 2012* The Fractalist. Memoir of a Scientific Maverick, *New York: Pantheon Books, Mandelbrot proclaims that "What a contrast [between 2012, or even 1986] with the period around 1960! Then Levy stability [for extreme value distributions] was viewed as a specialized and uninteresting concept.")*

Abstract The three preceding theories can only manage stable distributions; this chapter looks into the issue of unstable distributions. For stable distributions, risk estimates and their confidence intervals can be assumed to merge at infinity. However, for "unstable" distributions, those with infinite variances, risk estimates may "wobble." This chapter will therefore address the key question of bridging the gap between the infinite population assumed by probability and statistics with the finite samples of experience. This chapter contains a discussion of catastrophe ("stability") measures and their relationship to power laws. Power laws are a simplified means to show how estimating risk to systems can be cognitively dangerous if only because the systems themselves as studied can be very "fragile," or unpredictable. So, while it may be obvious that in science various "shocks" may be very challenging to predict, it may also be true that nonlinear behavior in systems can make evaluation of risk to these systems cognitively challenging, if not dangerous.

C.E. Taylor, *Robust Simulation for Mega-Risks*, DOI 10.1007/978-3-319-19413-4_5

5.1 Introduction: More Challenges to Tradition, Extreme Value Diagnostics, Power Laws, and the Wobble

One critical topic not discussed so far pertains to the treatment of *extreme value* distributions, such as nonlinear behavior, system shocks, and collateral events. The three theories that have been discussed up to this point can only manage stable distributions; therefore, a method is needed to deal with unstable distributions. In other words—how do we bridge the gap between the *infinite* population assumed by probability and statistics with the *finite* samples of experience? For "stable" distributions, those with finite variances and risk estimates and their confidence intervals can be assumed to merge at infinity. For "unstable" distributions, those with infinite variances, risk estimates may "wobble."

This chapter contains a discussion of catastrophe ("stability") measures and their relationship to power laws. Power laws are a simplified means to show how estimating risk to systems can be cognitively dangerous if only because the systems themselves as studied can be very "fragile" or unpredictable. So, while it may be obvious that in science various "shocks" may be very challenging to predict, it may also be true that nonlinear behavior in systems can make evaluation of risk to these systems cognitively challenging, if not dangerous.

This chapter uses a distinction by von Mises between probability and statistics, on the one hand, and the theory of distributions, on the other hand.[1] The theory of distributions can in some sense be termed a theory of mathematics, with an emphasis on parametric modeling and its consequences. The theory of probability and statistics instead can be nonparametric or can concern itself chiefly with the relationship between data samples, scientific information and theories, and parametric modeling.[2]

So far, in all three views of probability and statistics, there has been an implicit or explicit dependence on the law(s) of large numbers. These law(s) are not the entire story even within these theories. This chapter provides a simplified catastrophe index that assists in determining whether or not the law(s) of large numbers hold—at least as based on the sample to date. This chapter will show, for instance, that von Mises was overly circumspect when he limited his views to a collective that consisted chiefly of random and fairly inconsequential data without various types of perturbations. Various events can happen together as typically happens in disasters in which immediate food supplies may suffer immediately after a large-scale flood or a dam may fail during an earthquake and require a very rapid evacuation. These

[1] See p. 99 in Von Mises, Richard, 1957, *Probability, Statistics and Truth*, New York: Dover Publications, Inc.

[2] The use of three types of distributions in, for instance, Gumbel, E. J., 1958, *Statistics of Extremes*, Mineola, N. Y.: Dover Publications, Inc. (republished in 2004), illustrates how a smoothing approach has arisen for the theory of extremes and may be considered as part of the theory of distributions and less a part of the variations of how variable extreme value and other distributions may become. This variability is stressed in many of the charts in this chapter.

types of events or processes are not included in "collectives" as defined by von Mises, but one may still find that these events are part of processes for which one may qualifiedly say the law(s) of large numbers and central limit theorems still hold. The function of the catastrophe index in this chapter is to distinguish between distributions with finite variance and distributions with infinite variance, the latter called "fat tailed." A finite variance is all that is required for the law(s) of large numbers and the central limit to hold. From the standpoint of simulations, the cat index(es) provided in this chapter assist—better than the coefficient of skewness—in estimating how many simulations should be used to achieve conventional statistical stability.[3]

However, there can be applications of these laws when some of the data are very consequential. At the same time, there is a realm that this index can indicate in which the law(s) may not hold—at least according to the data at hand. For example, von Mises was aware that it is "of course possible for highly improbable accumulations of calls to happen on the same day." This comment today is commonplace when one considers mega-data and mega-hits on the Internet, phones, or electric power needs for immensely popular activities or for such websites that require enormous numbers of users or for special occasions or applications of computers or air conditioning.[4] A great many phenomena over and above many electrical engineering applications have extreme or almost extreme distributions: floods, Great Depression, Great Recession, earthquake, hurricanes, conflagrations, winter storms, pandemics, mega-construction risks, and species extinctions, to mention a few. The proposed index can be applied to distribution of wealth, distribution of values of properties in a portfolio, and other distributions in which the index is applied to positive nonzero values (e.g., distributions of net losses may need some added number to overcome events in which net losses are negative).

This simplified index has parallels with power laws that are often applied to systems. An addendum is included to explain why such simple models are used as diagnostics. A second addendum provides a discussion from chaos theory as to nonlinear systems for which extreme trajectories can be traced.

---

[3]The coefficient of skewness has multiple definitions, but in this case refers to Pearson's third central moment, a measure of the asymmetry of a distribution.

[4]In his 2006 book entitled *Noise*, New York: Viking, and, for instance, in Kosko, Bart, and Sanya Mitaim, 2004, "Robust stochastic resonance for simple threshold neurons," *Physical Review* E 70, 031911, Bart Kosko first shows that the law(s) of large numbers and associated central limit theorem encounter problems when faced with many chaotic systems, "impulsive" Cauchy noise, and the "bursty" rate of Ethernet traffic. He uses, for instance, what are called alpha-stable distributions as samples of extreme value distributions that can provide useful indexes similar to the catastrophe indexes discussed in this paper (see many pages in *Noise*.) One question that these chapters do not address is whether or not the "wobble," "wiggle," or "burst" in these distributions pose very serious problems for estimating more distant fractile estimates as one constructs alpha-stable distributions for these extreme circumstances.

5.2 The One-Parameter Pareto Distribution for Assessing the Extremity of Distributions

There are several treatments of this distribution that are somewhat or very different from that presented here.[5]The Pareto distribution has long been used in catastrophe reinsurance (insuring insurers) as a means to capture and cover the extreme loss values possible in catastrophes. Taylor et al. have maintained that the use of the Pareto distribution in such pricing yields prices that are far too high at extreme loss levels.[6] Nonetheless, the simplified Pareto distribution is a one-parameter distribution with a probability distribution function[7]

$$F(x) = 1 - x^{-c} \text{ and}$$

a density function of

$$f(x) = cx^{-c-1} \text{ for } x \geq 1$$

Where "c" is the slope parameter (or "slope"), the mean is $c/(c-1)$ when $c>1$.

The variance is $c/[(c-1)^2(c-2)]$ when $c>2$.

To simulate a one-parameter Pareto distribution, use the following:

1. Pick U_i in $U(0,1)$.
2. Derive $X_i = (1/Ui)^{1/c}$.

where U is a uniform distribution on the unit interval (0, 1).

In the above simulation process, the random numbers should be considered to comprise ultimately as being a "rectangular distribution," one comprising a straight line when ordered from low to high. Alternatively, one can define the following implicitly rectangular distribution for a sequence of N trials:

$$1/2N; 1/2N + 1/N; \ldots; 1/2N + (N-1)/N$$

Chart 5.1 below shows normalized Pareto variants resulting from the extreme values of c of 0.5, 1.0, 2.0, and 5.0 of 1000 simulations in the sense that the maximum loss is 1.0 (Y-axis). The slope 2.0 is used as the boundary for "heavy-tailed"

[5] See Embrechts, Paul, Claudia Klueppelberg, and Thomas Mikosch, 2003, *Modelling Extremal Events*, Berlin: Springer, fourth printing, p. 162; Gumbel, E. J., 1958, *Statistics of Extremes*, Mineola, N. Y.: Dover Publications, Inc. (republished in 2004), p. 45; and Law, A. M. and Kelton, W. D., 1991, *Simulation Modeling and Analysis*, New York: McGraw-Hill, p. 413.

[6] Taylor, C., J. Lemaire and C. Tillman (1994), "A New Earthquake Insurance and Reinsurance Index: Uncertainties and Future Developments," *Uncertainty Modeling and Analysis: Theory and Applications*, B. M. Ayyub and M. M. Gupta, (eds.), 1997. Elsevier Science BV., pp. 497–514.

[7] The treatment here comes from Hastings, N. A. J. And J. B. Peacock, 1975, *Statistical Distributions*, London: Butterworths, pp. 102ff.

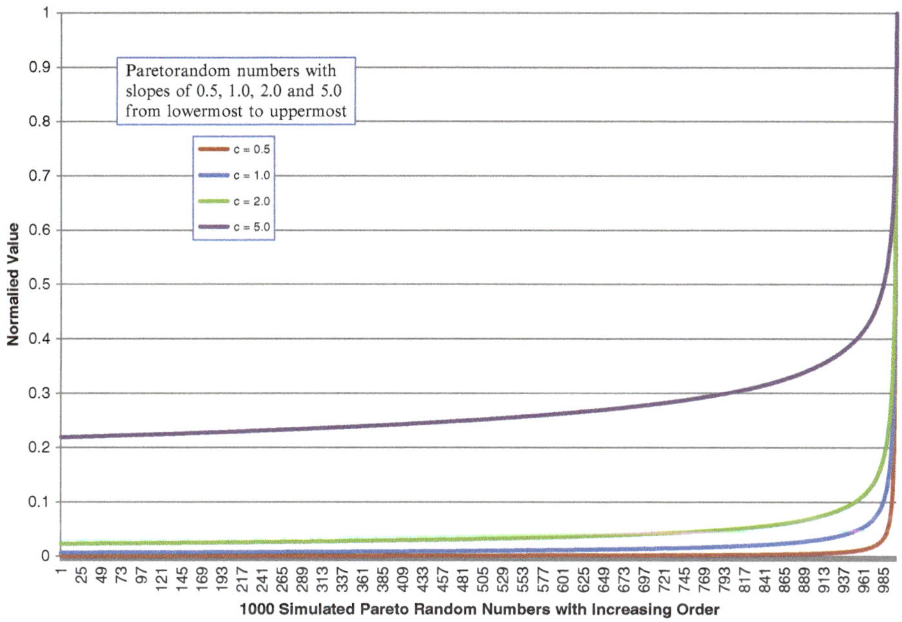

Chart 5.1 Pareto pdfs with slopes of 0.5, 1.0, 2.0, and 5.0 from uppermost to lowermost

distributions.[8] Note that catastrophe reinsurance pricing has in the past often postulated a "*c*" value well below 1.0. For this case, the arithmetic mean value is infinite or undefined. When the arithmetic mean value is undefined, there are major issues the strong law of large numbers does not hold nor does its correlate the central limit theorem.[9] Note that a power law is a straight line in log-log plot.[11] This log-log plot is shown in Chart 5.2.

Charts 5.1 and 5.2 apply to how "*c*" is derived by the above methods especially for low-probability and high-consequence risks. These include earthquake, hurricane, flood, and many other risks. However, there are risks in which the consequences can be very high but the probabilities also are more than "low." These medium-probability high-consequence events cover, for instance, new ventures, new lines of business, mega-construction projects, and very frequent occurrences of a natural disaster in a given locale. Charts 5.3 and 5.4 provide some insight into how

---

[8] The Pareto distribution is only one example of a fat-tailed distribution (when $c<2$). Source: Dr. Robert Riehemann letter December 22, 2014.

[9] According to p. 219, Mandelbrot, 2012, op. cit., "The 1900 [Bachelier's Gaussian] theory assumed that price jumps can be neglected—the mathematical concept being that 'prices vary continuously'…My 1962 counter theory allowed for discontinuities…" In effect, Mandelbrot was noticing with the more primitive computer outputs at the time that financial series may not comply with the strong law of large numbers.

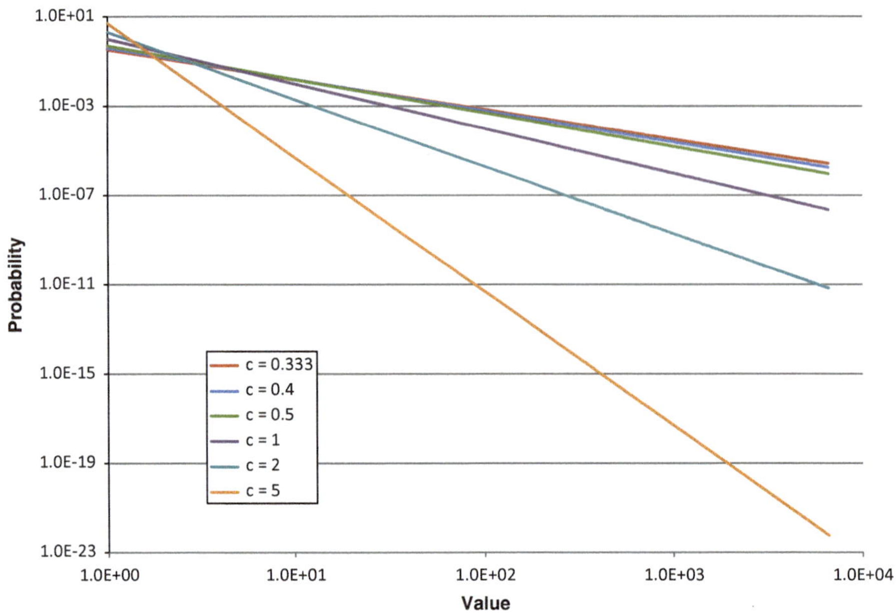

Chart 5.2 ln-ln plots of Pareto distributions with slopes (from lowest on the right to highest) of 0.333, 0.4, 0.5, 1, 2, and 5

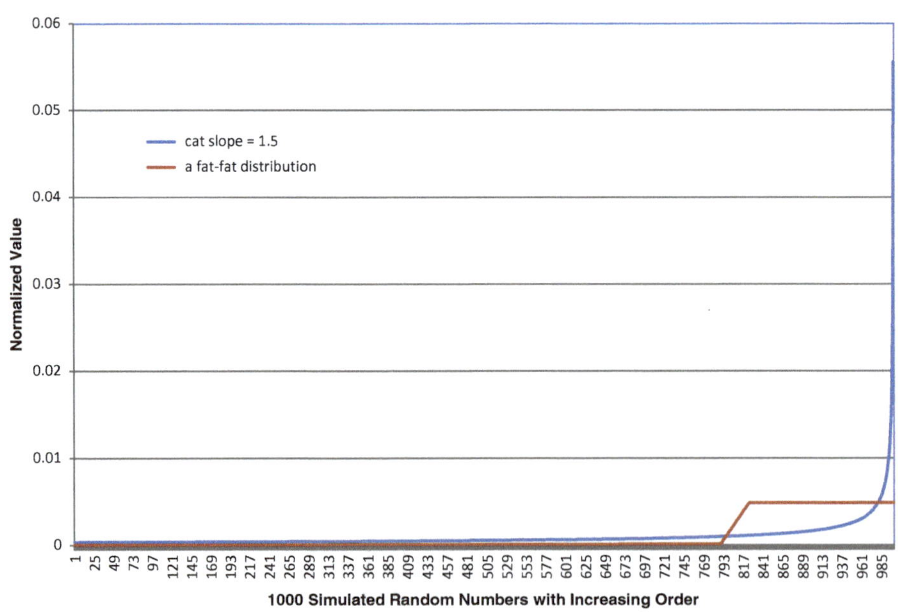

Chart 5.3 Contrast between a normalized extreme value distribution (cat slope = 1.5) and a fat-fat normalized distribution (fatness starts at 80th centile and plateaus, cat slope = 1.55)

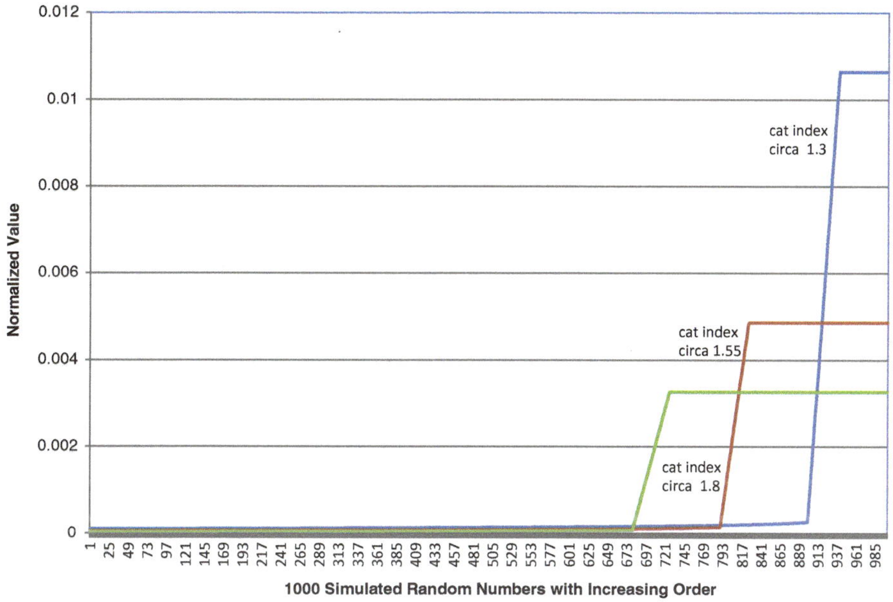

Chart 5.4 Illustration of normalized high-consequence distributions with varying probabilities (from somewhat low, 10 %, to higher (20 %) to even higher (30 %)

well these "heavy-tailed" but "medium consequence" phenomena may be evaluated in terms of the index provided in this chapter.

As in Chart 5.1, both Charts 5.3 and 5.4 are normalized Pareto variants from robust simulations.

In Chart 5.3, one of the curves represents a more typical low-probability high-consequence model in which the extreme tail dominates the overall result. Its slope is 1.5. In contrast, the second "curve" represents a risk in which the probability of total loss becomes very high as this probability approaches 20 % (or 80 % chance of non-exceedance). The normalized curves look very different, but oddly enough the slope of the second is slightly above 1.5.

Chart 5.4 shows normalized accounts of such higher-probability high-consequence "curves" as the probability of a high-consequence event increases. As this occurs, then the slope is reduced. If the high consequence consumes about 10 % of the loss distribution, then the slope calculated is about 1.3, as in Chart 5.4; as the high consequence consumes about 20 % of the loss distribution, then the slope calculated is slightly above 1.5; as the high consequence consumes about 30 % of the loss distribution, then the slope calculated is about 1.8. In general, one will find different slopes or cat indexes depending as well on the probability of occurrence. High-probability high-consequence events, activities, or processes represent situations in which these high consequences (sometimes ruinous) become almost or actually expected. The slope used here as a cat index does not cover very well, for instance, cases in which the high-consequence events, activities, or processes are almost certain.

5.3 A Simplified Account of Qualitative and Quantitative Approaches to Systems Evaluations

Very qualitative accounts of system reliability are common. One example is the field of ecology where vast numbers of studies have been undertaken both before and after Rachel Carson's *Silent Spring*.[10] Carson discusses many instances of how insecticides have damaged habitats. One of many such current issues pertains to how DDT has damaged the habitat off the Palos Verdes coast, in the intensely studied Palos Verdes shelf.

One account of the origins of damages points out many negative consequences. For instance, five decades of DDT and PCBs have been discharged through wastewater pipes. PCB was discharge until 1971. One view of DDT is that it does not breakdown and is easily transported from sediment into the water columns. Brown pelicans, peregrine falcons, and bald eagles had their reproductive cycles disrupted as when their egg shells became too thin. Bottom feeding fish ceased to be edible.[11]

Another account stresses how clean-up efforts and consequences of this DDT have not been readily predictable. Ninety percent of the DDT off the Palos Verdes Peninsula had vanished in a surprisingly short 5-year period after many years of slow, gradual decline. One expert noted: "It doesn't make sense to me that this degree of a change would have occurred within the last 5 years. It's very difficult to assess where it went."[12]

These more qualitative efforts depend heavily on instrumented values in order to develop projections, and in this case projections have not proved to be as severe as expected. In addition, there are competing organizational and legal interests involved, so that it becomes even more incumbent on those making estimates and updating them to be as impartial as possible.

Infrastructure and other systems are continually being studied. In addition to more qualitative studies buttressed by much data, there are three approaches to systems, and each of these approaches can be carried out in varying detail relative to various hazards that may impact the system in question:

1. Boolean or connectivity approaches: these treat system nodes and components as being either fully functional or fully disabled. This approach greatly simplifies how systems can be evaluated. To achieve this approach probabilistically, one must determine the probability level at which nodes and components are regarded as failing, and one must have fragility models that distinguish only between success and failures of nodes.

[10] Carson, Rachel, 1962, *Silent Spring*, Boston: Houghton Mifflin Company. The influence of this work and others on the view of science and its applications is discussed in Chap. 9.

[11] "DDT/PCBs Off the Shores of Palos Verdes," in *palos verdes.com/eco/*political/ddt, Accessed June 24, 2013. Accounts such as these are traditionally disputed because they are part of court disputes and potential settlements.

[12] Cone, Marla and *Environmental Health*, 2013, "The Mystery of the Vanishing DDT in the Ocean Near Los Angeles," *Scientific American*, March 13, accessed June 24, 2013.

2. Capacitated approaches: these treat system nodes and components as having various levels of capacity. The entire system can be evaluated in terms of total demand desired and the capacity available to achieve this desired demand. For instance, one may treat a roadway segment as needing to carry on peak situations 10,000 vehicles. The capacity evaluation would then determine whether or not under varying circumstances these 10,000 vehicles could run on the roadway segment in question. Capacitated approaches permit the evaluation of nodes and components in terms of various levels of capacity or functionality.
3. Flow approaches: these treat the system as having various physical properties connecting nodes and links, along with various demands. For instance, water systems may require adequate pressures along with connectivity so that demands may be met. These evaluations can be carried out at extremely detailed levels of analysis.

Discussed next, power laws turn out to be one way in order to provide a very broad quantitative view of system reliability.

5.4 A Simplified Account of the Power Law in Relation to the One-Parameter Pareto Distribution

The account here follows Barabasi,[13] who is interested in applying the power law to systems. In particular, systems comprised primarily of a hub and spoke—in which the hub has many connections and the spokes each have few—can follow an extreme distribution. These systems have many nodes whose damage can yield only local damage to the overall system. Yet, damage to the hub (or a very few nodes) can have severe repercussions on the system. This discussion also shows the parallels between power laws and the one-parameter Pareto distribution.

Modifying Barabasi slightly, the power law has the form:

$$Y = bX^{-\text{alpha}}$$

in which

Y = the fractional result (e.g., the fraction of nodes with X connections to other nodes)
b = a scalar
alpha = the slope

For Barabasi, X is an integer. Here, $X > 0$. Also for Barabasi, alpha ranges from 2 to 3. These values, it turns out, are for what he calls a "scale-free" network. A scale-free network is one governed by power laws such as the one above. In effect, the ranges from 2 to 3 result in very extreme values. The relation between the Pareto distribution slope of "c" in the previous subsection and the "alpha" here is:

$$\text{alpha} = 1/c.$$

[13] Barabasi, Ibid., pp. 74, 85.

That is, for ranges of alpha from 2 to 3, "*c*" in the Pareto distribution ranges from 0.333 to 0.5, extraordinarily low values resulting in very extreme distributions. Needless to say, one can have less extreme values of "alpha" in the above and still have the equivalent of a non-Gaussian distribution (with a Pareto slope of 2 or less).

As with the Pareto distribution, one may use a ln-ln plot for the power law. For Chart 5.2, the values of alpha would be 3, 2.5, 2, 1, 0.5, and 0.2 from lowest to highest on the right side.

Note that Barabasi claims that power laws apply to "phase transitions," such as when something turns from a liquid to a solid or something becomes magnetized (with atoms now spinning in the same direction).

5.5 Challenges of the One-Parameter Pareto Distribution and Power Laws to Traditional Theories

This approach departs to some extent from the approach that is taken in, for instance, the important work of E. J. Gumbel, who is largely concerned with a distributional approach rather than what von Mises calls a statistical approach. For the approach being used here, the largest value of a sample would be too eccentric to employ, whereas the use of the 99th centile estimate can provide "mass" to the estimate in the following sense: the "shape" or "slope(s)" of the 99th centile mass are not predetermined. Thus, the "largest" value can be below what this 99th centile test would estimate, or it could be well above this 99th centile estimate would estimate. Shapes of distributions are not presupposed on the account being presented here. In addition, this approach is clearly empirical in the sense that with further samples the results may change. What seems to be an unstable distribution may eventually be seen to be a stable one and vice versa. However, the work of E. J. Gumbel on the Pareto distribution may in some or many cases correspond to the estimation procedure used here.[14]

The following rules apply to the extent that this test applies:

1. If the Pareto slope is 2 or less, then the statistical variance (i.e., the standard deviation) is infinite.
2. If the Pareto slope is 1 or less, then the statistical mean is infinite.
3. If the Pareto slope is 2 or less, and especially if the Pareto slope is 1 or less, an estimation may be right 99 % of the time and very wrong 1 % of the time.

[14] Gumbel, E. J., 1958, *Statistics of Extremes*, Mineola, N. Y.: Dover Publications, Inc. (republished in 2004). Note that Gumbel stresses finding the largest value rather than using a "mass" estimate such as the 99th centile estimate. Ultimately it appears that there may be convergence between Gumbel's approach and that used here with respect to mega-simulations. The author also agrees that for extreme value distributions, the mode is extremely important and that one cannot resort merely to normal or lognormal distributions (see p. 345 in Gumbel).

One can translate these into power laws as defined above:

1a. If the power law is 0.5 or above, then the statistical variance is infinite.
2a. If the power law is 1.0 or above, then the statistical mean is infinite.

These findings undermine the traditional (Gaussian) theories of statistics that rely heavily on the strong law of large numbers for convergence of values should the number of samples proceed to infinity. That is, if the Pareto slope is 2 or less, then values of "fractiles" (e.g., 100-year losses) are ill-defined—no matter how many samples are used. If the Pareto slope is 1 or less, not even the mean values are well-defined.

Thus, the presence of extreme values undermines the assumption that—even if the number of samples were virtually infinite—for all distributions there must exist unique statistical or probabilistic estimates. Taleb regards various Black Swan events as being "incomputable." In this essay, specific extreme value distributions will exhibit a "wobble," the horizon of a permanent lack of uniqueness for some value or set of values.[15]

Chart 5.5 illustrates how the wobble occurs for Pareto slope values of 1.5 and especially 0.5 but not for the Pareto slope value of 2.5. In these cases, 1000 uniform random variates are used to derive these outcomes, and the final mean value is forced to be 1/1000 or 0.001. In actuality, a "true" Pareto with a slope of 0.5 has no "final" value. Instead, it continues to "wobble" indefinitely.

Chart 5.6 illustrates how the wobble occurs for estimates of the standard deviation relative to the "final" mean values. Again, for the Pareto slope value of 0.5, not only is the "final" mean illusory but so is its alleged standard deviation. For the Pareto slope value of 1.5, the standard deviation "wobbles." Eventually, for a Pareto slope value of 2.5 (the same as the slope value for the exponential distribution), the standard deviation stabilizes.

---

[15] From p. 138, Taleb, Nassim Nicholas, 2012, *Antifragile: Things that Gain from Disorder*, New York: Random House. If as in Chap. 6 deviations are used instead of standard deviations in order to define extreme distributions and if this means that average deviations are used, then one can derive from these a mean value of the distribution. Thus, if the catastrophe index here is 1 or less, deviations cannot be used to redefine the distribution. Nolan's work, for instance, stops in the development of extreme distributions when the catastrophe index is 1 or below: see Nolan, John P., 2009, *Stable Distributions: Models for Heavy Tailed Data*, accessed on the Internet February 27, 2013. Years after the term "wobble" was selected, I found the term "wiggly" on p. 219 of Mandelbrot, Benoit B., 2012, *The Fractalist: Memoir of a Scientific Maverick*, New York: Pantheon Books: "All price charts...are qually wiggly. "Wiggly" is hardly a scientific term—[b]ut that is exactly what we see in the [cotton price change] data: a fractal pattern.

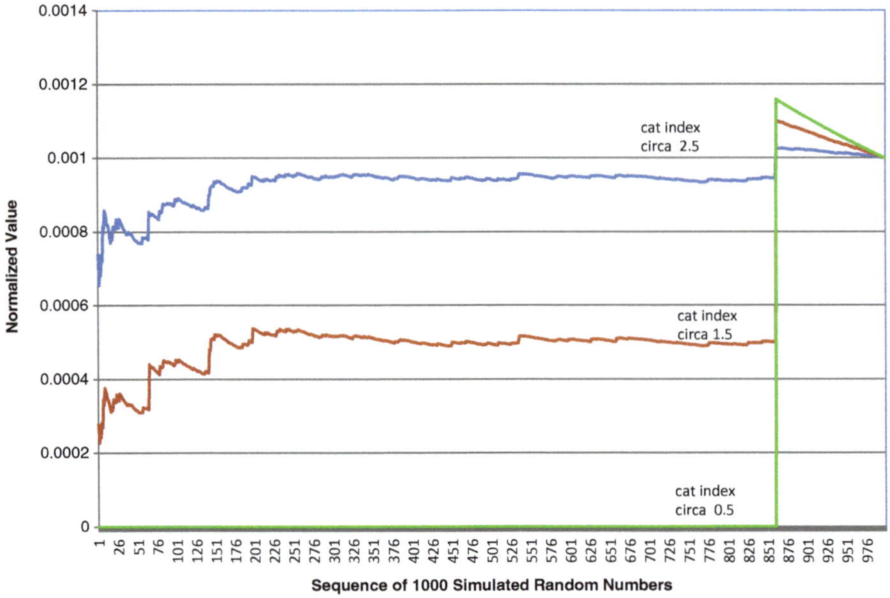

Chart 5.5 Comparison of running mean values for three Pareto slopes: 2.5, 1.5, and 0.5 from top to bottom with sharp incline at 868th simulation (end value of 0.001 for the mean results because 0.001 is the mean value postulated for each 1000 simulations for each slope)

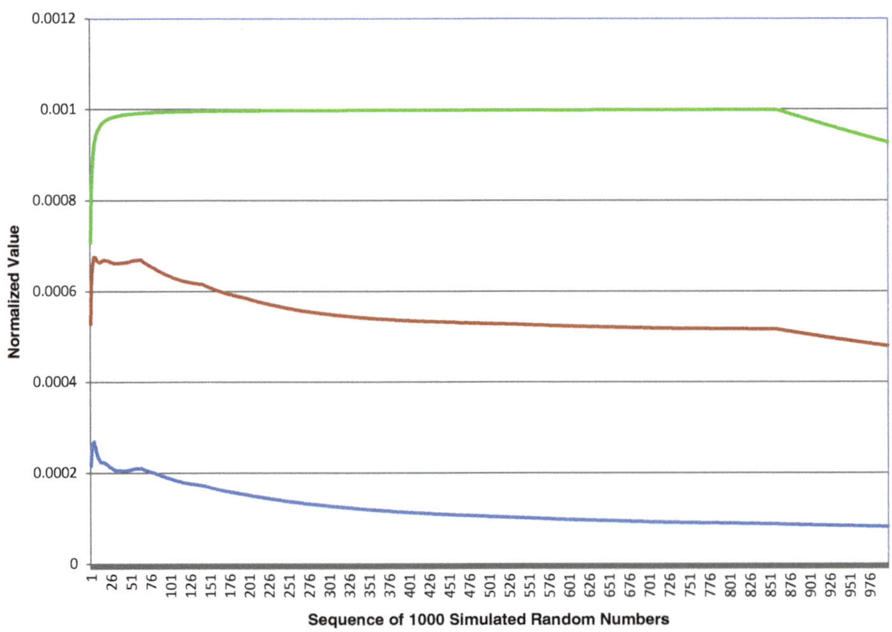

Chart 5.6 The "Wobble" in standard deviations lower than 2.0 and stability in standard deviations above 2.0 (running standard deviations relative to three Pareto slopes of 0.5, 2.5, and 1.5 from *top* to *bottom* and relative to final mean value)

5.5.1 *Even the Lognormal Distribution Can Become Extreme*

Oddly as it may seem, the lognormal distribution is not per se oblivious to more heavy-tailed considerations, some of which imply that the central limit theorem and the strong law of large numbers do not apply in specific cases. As David Perkins has indicated,[16] one may view the mean value as being

$$\mu = (1/x) * ([\ln x - \ln \mu] / \ln \sigma)^2$$

Then, as the coefficient of variation ($cv = \sigma/\mu$) increases to 4, the tail of the distribution approaches a scalar times 1/x, which is known to diverge. Augmenting Perkins's remarks, one finds further results using the catastrophe index in Chap. 5 along with 1000 balanced uniform samples. In particular, a cv of 1.6 may be close to or below 2, a cv of 7.6 is very close to 1, and a cv of 12 is well below 1. Using these finite-sample tests, as the cv approaches 1.6, then "variance" may be unstable since the cat index is close to or below 2, and as the cv approaches 8, even the mean value may be unstable since the cat index is close to or below 1. So, even the lognormal distribution cannot automatically be regarded as being subject to Gaussian theorems

5.5.2 *But There Can Be Lots of Perturbations in Stable Distributions*

In the application of the cat index to actual loss distributions, most have been at least barely statistically stable, that is, with cat indexes above 2. Below are two illustrations. Chart 5.7 provides the example of a loss distribution that has a cat index of 7.9. This, though, implies that the distribution is clearly not normal. Chart 5.8 provides an example of a loss distribution that has a cat index of 2.5. Curiously enough, this is the cat index for an exponential distribution (not an exponential type distribution in Gumbel's sense). However, the shape of the distribution is by no means exponential.[17]

Thus, the extraordinary efforts that von Mises employs to assure that the law(s) of large numbers holds is excessive. Most actual distributions may be "stable." However, even stable distributions, such as that shown in Chart 5.8, can give rise to major issues.

---

[16] David Perkins, written comm., June 2, 2009.

[17] These charts are produced by joint work with Zhenghui Hu.

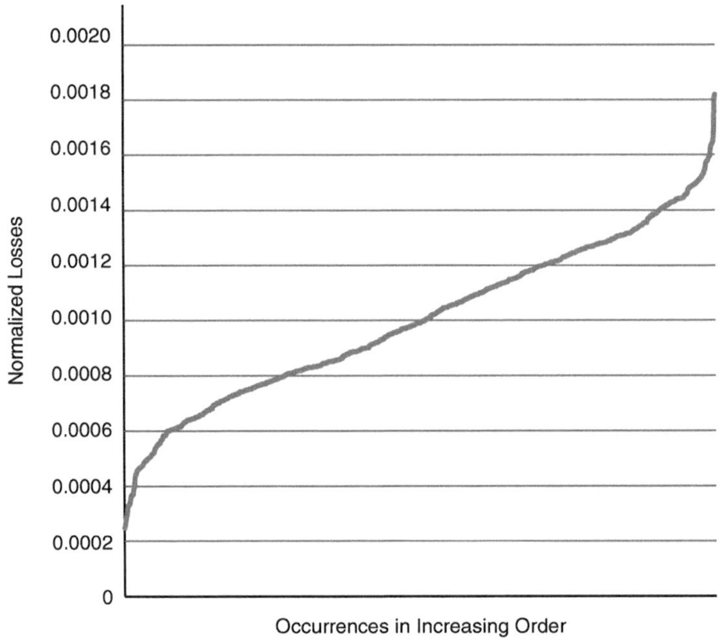

Chart 5.7 Example of a loss distribution with a cat index of 7.9

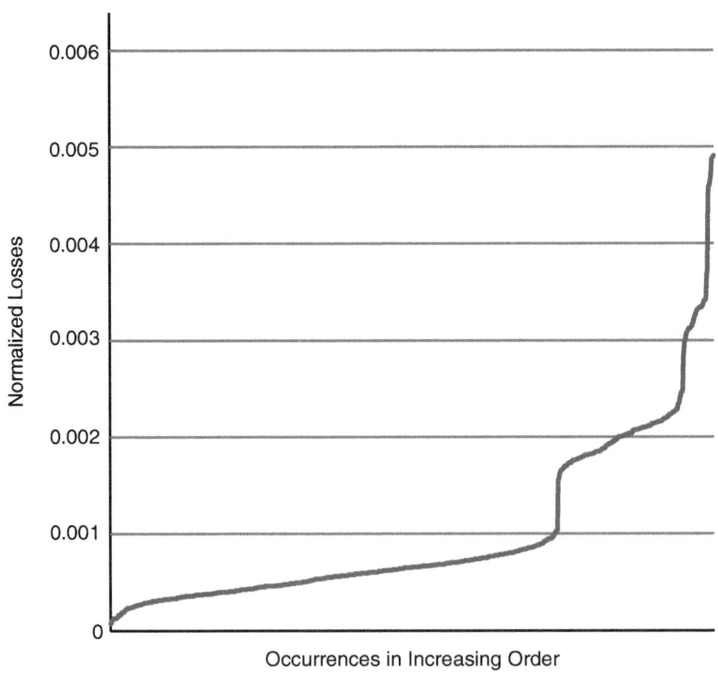

Chart 5.8 Example of a loss distribution with a cat index of 2.5

5.6 Summary of Extreme Value Diagnostics, Power Laws, and the Wobble

Simplified empirical tests can diagnose how "dangerous" the derived distributions turn out to be. For very "dangerous" distributions, with Pareto slopes less than 2, then many values in the resulting distribution may "wobble," or not be uniquely defined. If the Pareto slopes are 1 or less, then even the mean values may "wobble" or not be uniquely defined. Most distributions—even those for shocks to systems—may be statistically "stable." Some systems as well as distributions may illustrate unstable circumstances. In these circumstances, things may go well 99 % of the time but fall apart 1 % of the time. Hence, tests of distributions that only go so far may miss these extreme cases.

Hence, this section enhances the reasoning in the previous sections that relies heavily on the absence of any cases in which there have been an infinite number of samples. In some cases, even the idealized presence of infinitely many samples will not result in uniquely defined probability or statistical values.

Addendum 1: Why Such Simple Extreme Value Models Have Been Deployed

Many of the treatments of extreme value distributions are mathematical in the sense that mathematical or parametric models are presumed, and the user feels considerable control over the resulting values. The treatment here diagnoses the dangers of existing distributions, as defined from finite data.

In the literature, a number of graphical measures are used to define whether or not a distribution is "heavy tailed." In cases here surveyed, "heavy-tailed" means having no finite statistical variance. In some cases surveyed, these tests may provide more information as is clearly the case with the diagnostic tests provided here. However, this greater information tends to be opaquely used because these tests are graphic; when exposed to actual data, their results are subject to interpretation.

For example, the Hill's estimator is one such test.[18] If one applies this to a Pareto distribution with a predefined slope, then the Hill's estimator assists in defining this slope, as in the following Chart 5.9 in which the predefined slope is 1.0.

However, as in Chart 5.10, the application of the one-parameter Pareto distribution to the exponential distribution (no matter what positive slope is used for this distribution) turns out to yield a Pareto slope of 2.5. Yet, as in Chart 5.10, Hill's

---

[18] Sources for Hill's estimator are Embrechts et al., op. cit., pp. 330ff.; Resnick, S., 2007, *Heavy-Tail Phenomena*, New York: Springer, pp. 80ff. and Beirkut, J. and G. Mattheys, 2000, "Quantile Estimation for Heavy-Tailed Data," 23/03/2000mistis.inrialpes.fr/work/extreme/jb.ppt.

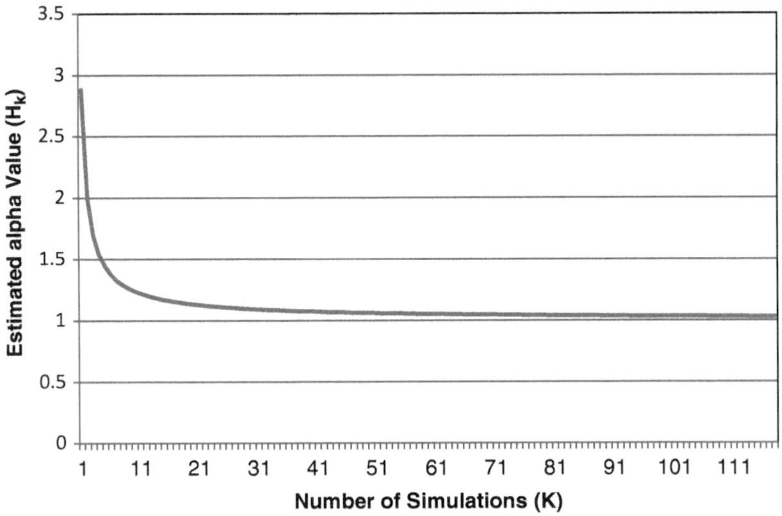

Chart 5.9 Hill's estimator, Pareto, alpha=1

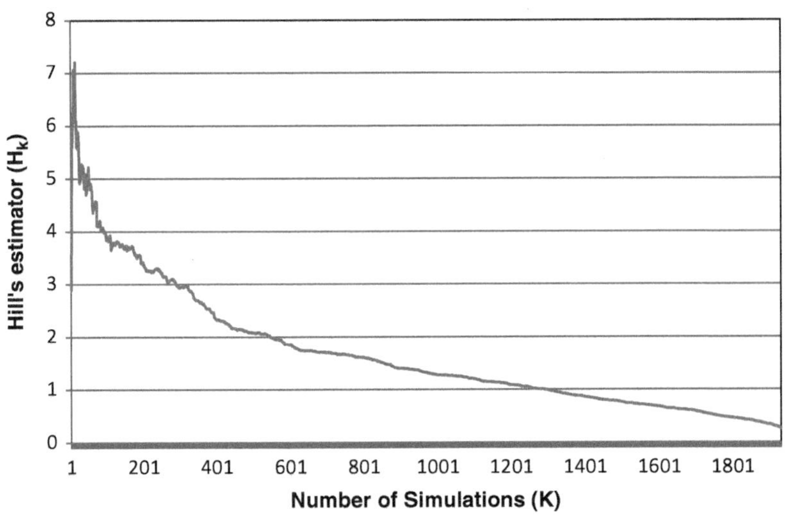

Chart 5.10 Hill's estimator, simulated exponential distribution

estimator provides no clear picture for the exponential distribution. And, as in Chart 5.11, when Hill's estimator is applied to an actual distribution, here one on historic losses as provided by T. Jagger,[19] then there is no unambiguous solution.

[19] T. Jagger, written comm., December 7.

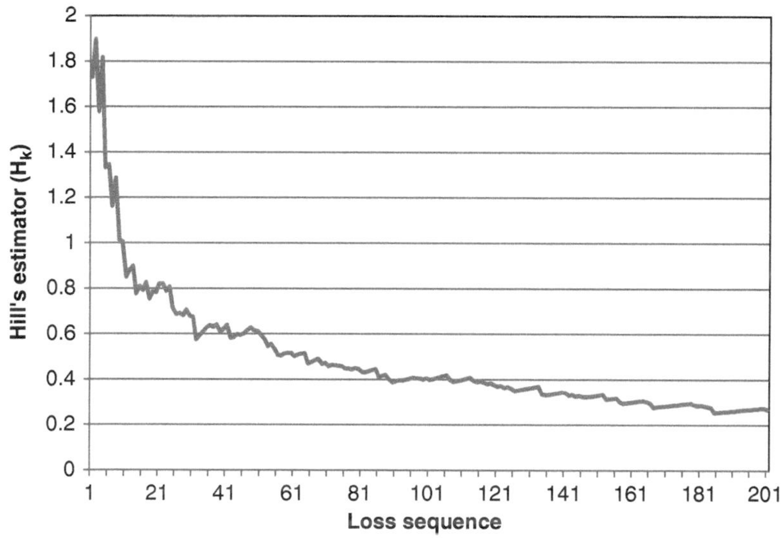

Chart 5.11 Hill's estimator, from T. Jagger, historic hurricane losses

Another well-known diagnostic test for heavy tails is Dekker's estimator.[20] Chart 5.12 shows how this applies to a one-parameter Pareto distribution with a slope of 1.0. There is a graphical patter in such a parametric case. However, in Chart 5.13, when applied to historic hurricane losses, the subsequent figure shows that ambiguities arise.

Addendum 2: Heavier-Tailed Distributions and Genuinely Nonlinear Models

Nonlinear equations are those that involve second-order and higher values (e.g., x**2). S. Strogatz characterizes nonlinear "systems" as follows:

> Linear systems can be broken down into parts. Then each part can be solved separately and finally recombined to get the answer... This... underlies such methods as normal modes, Laplace transforms, superposition arguments, and Fourier analysis...a linear system is precisely equal to the sum of its parts...Whenever parts of a system interfere, or cooperate, or compete, there are nonlinear interactions going on.[21]

[20] Sources as for the Pickand's estimator are Embrechts et al., op. cit., pp. 328ff; Resnick, S., pp. 90ff. and Beirkut, J. and G. Matthys, op. cit. Resnick is especially critical of the use of this estimator and develops two modified versions of this estimator.

[21] From Strogatz, Steven H., 1994, *Nonlinear Dynamics and Chaos: With Applications to Physics, Biology, Chemistry, and Engineering*, Cambridge, MA: Perseus Books Publishing, LLC, pp. 8, 9.

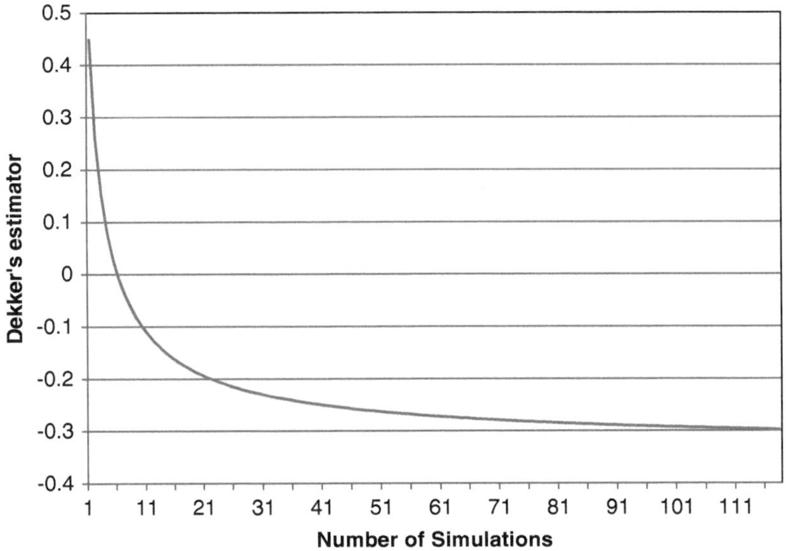

Chart 5.12 Dekker's estimator, Pareto, alpha=1

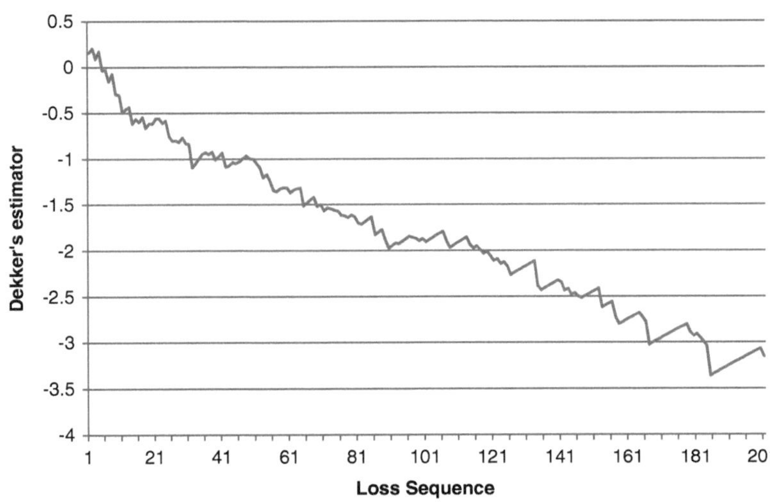

Chart 5.13 Dekker's estimator, Tom Jagger, historic hurricane losses

For example, Strogatz defines the following two-dimensional linear system of the form:

$$Dx / dt = ax + by$$

$$Dy / dt = cx + dy$$

In which x and y are coordinates, t is time, and a, b, c, and d are "parameters" (i.e., coefficients to be determined).[22]

In some cases, statistical methods can effectively linearize some formally nonlinear equations. However, for genuinely nonlinear problems, Menke maintains:

> There are no simple means for deciding whether a non-linear inverse problem has a unique solution that minimizes prediction error in the absence of prior information…[23]

Cambel maintains that:

> there are no explicitly general solutions to linearize mathematical problems. In the past, there was a tendency to deal with nonlinearity by consideration such problems as aberrations and ignoring them. With increasing populations, dwindling resources, and rising Expectations, we can no longer indulge in this cavalier attitude.[24]

Thus, ignoring the many models used in complex systems evaluations, even single models may pose serious issues for derive unique estimates. The use of "prior" information permits of course alternative approaches with alternative outcomes. So, too, diverse investigators may use different tests, forms, parameters, and systems approaches (combining different models). At the same time, the prior assumption of normality may too yield fallible or erroneous conclusions.

As regards statistical forecasting, N. Silver discusses nonlinear modeling to emphasize current weather forecasts in which small changes are critical, how things can behave in strange and mysterious ways in which they interact with each other (the theory of complexity), and how in the atmosphere the dynamic memory of the atmosphere can erase itself over time.[25]

References

Ayyub, B. M., & Gupta, M. M. (Eds.). (1997). *Uncertainty analysis in engineering and the sciences: Fuzzy logic, statistics and neural network approach*. Dordrecht: Kluwer Academic.

Beirkut, J., & Mattheys, G. (2000). Quantile estimation for heavy-tailed data. mistis.inrialpes.fr/work/extreme/jb.ppt

Benoit, B. (1983). *The fractal geometry of nature*. New York: W. H. Freeman and Company.

Benoit, B. (2012). *The fractalist: Memoir of a scientific maverick*. New York, Vintage.

Cambel, A. B. (1993). *Applied chaos theory: A paradigm for complexity*. San Diego: Academic.

Carson, R. (1962). *Silent spring*. Boston: Houghton Mifflin Company.

Cone, M., & Environmental Health. (2013). The mystery of the vanishing DDT in the Ocean near Los Angeles, scientificamerican.com, March 13, 2013.

---

[22] From Strogatz, Ibid., p. 123.

[23] From Menke, W., 1989, *Geophysical Data Analysis: Discrete Inverse Theory*, San Diego: Academic Press, p. 153.

[24] From Cambel, A. B., 1993, *Applied Chaos Theory: A Paradigm for Complexity*, San Diego CA: Academic Press, p. 9.

[25] From Silver, Nate, 2012, *the signal and the noise: why so many predictions fail—but some don't*, New York: the Penguin Press, pp. 118, 120, 124, 132, and 172.

Embrechts, P., Klueppelberg, C., & Mikosch, T. (2003). *Modelling extremal events*. Berlin: Springer, fourth printing.

Fisher, R. A. (1944). *Statistical methods for research workers* (9th ed.). London: Oliver and Boyd Ltd.

Gumbel, E. J. (1958). *Statistics of extremes*. Mineola: Dover Publications.

Hastings, N. A. J., & Peacock, J. B. (1975). *Statistical distributions*. London: Butterworths.

Kosko, B. (2006) book entitled *Noise*. New York: Viking.

Kosko, B., & Mitaim, S. (2004). Robust stochastic resonance for simple threshold neurons. *Physical Review E, 7*, 1–10.

Law, A. M., & Kelton, W. D. (1991). *Simulation modeling and analysis*. New York: McGraw-Hill.

Mandelbrot, B. B. (2012). *The fractalist: Memoir of a scientific maverick*. New York: Pantheon Books.

Menke, W. (1989). *Geophysical data analysis: Discrete inverse theory*. San Diego: Academic.

Mises, V. R. (1957). *Probability, statistics and truth*. New York: Dover Publications.

Nolan, J. P. (2009). *Stable distributions: Models for heavy tailed data*.

Resnick, S. (2007). *Heavy-tail phenomena*. New York: Springer.

Resnick, S., Beirkut, J., & Matthys, G. op. cit.

Silver, N. (2012). *The signal and the noise: Why so many predictions fail—But some don't*. New York: The Penguin Press.

Strogatz, S. H. (1994). *Nonlinear dynamics and chaos: With applications to physics, biology, chemistry, and engineering*. Cambridge, MA: Perseus Books Publishing, LLC.

Taleb, N. N. (2012). *Antifragile: Things that gain from disorder*. New York: Random House.

Taylor, C., Lemaire, J., & Tillman, C. (1994). A new earthquake insurance and reinsurance index: Uncertainties and future developments,

Chapter 6
Battling Inductivist vs. Deductivist Theories (1900s to Present)

> *Coolly considered, this is a preposterous claim, which would have been universally rejected long ago, if those who made it had not successfully concealed themselves from the eyes of common sense in a maze of mathematics. (From pp. 388–389 in Keynes, John Maynard, 1921,* A Treatise on Probability, *London: MacMillan and Co.)*

> *I should like to say: mathematics is a motley of techniques of proof.—And upon this is based its manifold applicability and its importance. (p. 84c in Wittgenstein, Ludwig, 1967,* Remarks on the Foundations of Mathematics, *Cambridge, MA: the M. I. T. Press, first published in 1956, edited by G. H. von Wright, R. Rhees, and G. E. M. Anscombe and translated by G. E. M. Anscombe)*

> *George Cantor claimed that* the essence of mathematics lies in its freedom. *But mathematicians do not pick problems from the air for the pleasure of solving them. ...Do I claim that everything that is not smooth is fractal? That fractals suffice to solve every problem of science? Not in the least. (From pp. 178 and 299 in Mandelbrot, Benoit B., 2012,* The Fractalist: Memoir of a Scientific Maverick, *New York: Pantheon Books)*

Abstract This chapter provides an account of mathematical approaches that have been used in fitting models to distributions and in providing statistical acceptability tests of hypotheses. These helpful tools attempt to bridge the gap between finite samples and the infinite populations. Tools of interest in this chapter pertain to significance tests and fitting methods.

In terms of the underlying distributions needed for statistics, mathematicians in the past used binomial, Poisson, and normal (Gaussian) distributions. However, distributions have been expanded to include others such as the exponential distribution.

The theory of random probability distributions has been expanded so that indefinitely many distributions—far more than the 30, 40, or so that are familiar—can be constructed. This expansion of the theory of random probability distributions now permits investigators to postulate alternative underlying distributions with alternative solutions. This chapter stresses the multiple interpretations that can and should be derived from the use of statistics.

C.E. Taylor, *Robust Simulation for Mega-Risks*, DOI 10.1007/978-3-319-19413-4_6

6.1 Introduction to Mathematization of Statistics: Flexibility and Non-convergence

This chapter will cover the mathematization of statistics: tools developed to bridge the gap between finite samples and the infinite populations assumed by probability and statistical theories. The mathematization of statistics provides a cornucopia of valuable tools that are often being augmented. Tools of interest in this chapter pertain to significance tests and fitting methods. Such tools were first developed by statisticians at the turn of the century and are associated with such figures as K. Pearson and R. A. Fisher.[1]

The discussion of R.A. Fisher and related approaches is designed to address such questions as "does this mathematization require convergence (or at least "stability)?," "how can statistics have any value when standardized practices have resulted in so many errors?," and "do these methods bridge the gap between infinite populations and experience treated as finite samples?" As the chapter title suggests, these essays will ultimately emphasize how one should stress the power and flexibility of mathematical tools rather than their absolute certainty.

The theory of random probability distributions has been expanded so that indefinitely many distributions—far more than the 30, 40, or so that are familiar—can be constructed. This expansion of the theory of random probability distributions now permits investigators to postulate alternative underlying distributions with alternative solutions. The tests posited by R. A. Fisher and followers can be used as fallible diagnostics about whether or not the underlying distribution "washes out" with future data. The use of multiple underlying distributions belies R. A. Fisher's dictum that there is one correct solution.

The lack of unique outcomes can also be shown as one examines the flexibility that different investigators have in applying statistics. This is shown in inverse modeling. Also, similar problems of underfitting and overfitting apply in regression and other modeling.

6.2 Mathematical Tools Enhancing Statistical Practice: Flexibility or Incorrigibility

As has been stated previously in these essays, considerable mathematical attention has been placed on exponential, lognormal, and even more extreme distributions (e.g., the Weibull, Pareto, Gumbel, generalized extreme value, Cauchy). In contrast, the long traditions of Gaussian statistics have developed a potpourri of materials for

[1] These conflicts between K. Pearson and R. A. Fisher, on the one hand, and prominent Bayesians are discussed throughout McGrayne and Sharon Bertsch, 2011, *The theory that would not die: how Bayes' rule cracked the enigma code, hunted down Russian submarines & emerged triumphant from two centuries of controversy*, New Haven: Yale University Press.

any data for which normal distributions are assumed to fit.[2] R. A. Fisher alone is reputed to be the pioneer of randomization methods, sampling theory, tests of significance, maximum-likelihood estimation, analysis of variance, and experimental design methods. According to McGrayne, Fisher's p-values have been used millions of times. They are probability statements based on Gaussian assumptions. In addition, Neyman-Pearson theory of hypothesis tests have been enormously influential in applied mathematics.[3]

For frequency and Bayesian theories, in no cases have the tests used by themselves been verified by an infinite sample—even if the mathematical representation of findings may imply such an extension of the fit of a parametric equation. The parametric equations themselves are assumed or presupposed to apply and so by implication may be faced with "new cases" that show that they do not fit.

In contrast to the view that mathematics gains its power through its absolute certainty, one may maintain that the success of this mathematization—as in the very long-term rigorous development of the calculus or in the many mathematical tools for applications and statistics—resides in its benefitting these applications from standardized methods on which practitioners can rely. One should indeed be impressed by all of these developments. As extensive and valuable as these developments have been and continue to be, these though do not overcome the qualifications outlined in previous chapters. These developments and tools provide a flexibility to be discussed in this chapter.

This second view of mathematics is referenced in Chap. 2 with respect to comments by Herman Weyl or the treatment in Carl Boyer of the development of the calculus over more than two millennia. One finds similar findings in Bernoit Mandelbrot:

> Classical mathematics had its roots in the regular geometric structures of Euclid and the continuously evolving dynamics of Newton. Modern mathematics began with Cantor's set theory and Peano's space-filling curve. Historically, the revolution was forced by the discovery of mathematical structures that did not fit the patterns of Euclid and Newton. These new structures were regards…as "pathological"…as a 'gallery of monsters,' kin to the cubist painting and atonal music that were upsetting established standards of taste in the arts at about the same time. The mathematicians who created the monsters regarded them as important in showing that the world of pure mathematics contains a richness of possibilities going beyond the simple structures that they saw in Nature.[4]

This chapter begins with a contrasting view of mathematization that constrains itself to providing as much self-evidence as possible and later considers richer possibilities in the mathematization of probability and statistics.

[2] See, for instance, pp. 6–20 in Neter, John, William Wasserman, and Michael Kutner, 1985, *Applied Linear Statistical Models*, Homewood, IL: Irwin.

[3] See McGrayne, op. cit., pp. 47, 49, and 55.

[4] See p. 3 in Mandelbrot, Benoit B., 1983, *The Fractal Geometry of Nature*, New York: W. H. Freeman and Company, originally 1977.

6.3 Notorious Past Failures in Statistical Findings

The topic of incorrigibility in statistical practice yields a strong rebuke from cases in which statistical practice has come up with the wrong answers. A brief discussion of these alleged falsifications of statistical practice should arrest attempts to consider applications of mathematical statistics as being incorrigible.

The idiosyncratic work of N. Taleb points out failed statistical estimates in, for instance, economics, finance, and medicine. One of Taleb's hobby horses is iatrogenics—inadvertent adverse effects or complications resulting from medical treatment or advice. Among the many examples of iatrogenics that Taleb cites are replacement of butter by margarine, overuse of Vioxx, antidepressants in many cases, cortisone, hormone replacement theory, caesarean births, disinfectants, lobotomies, whitening of rice and wheat, and annual mammograms. In general, Taleb treats these errors as proof of extreme skepticism and the methodology of via negativa—progress through falsification or rejection of hypotheses since "justifying" truth is apparently more challenging than falsification. As incisive as are many of Taleb's insights into iatrogenics and elsewhere, the recourse to "T" or "F" is similar to the diatribe against medical research found, for instance, in *The Economist*.[5]

Taleb's explanations of these "failures" include the nonlinearity in the physiological response of biological systems, problem of scaling data for nonlinear models, preoccupation of data accumulation when nature has provided tens of thousands of years of data, accumulation of data yielding false data and more correlations, and lack of compatibility for "Black Swan" (extreme value) situations.[6]

In a similar vein, the late Berkeley statistician David Freedman has consistently reviewed various statistical findings and methods, including examples from medicine. For but one instance, Freedman and Petitti critique the salt hypothesis that higher levels of salt lead to higher levels of blood pressure so that salt intake should be cut by a factor of two or more. Freedman and Petitti find multiple interpretations of studies and suggest that with good diet, salt has almost no impact on systolic pressure.[7]

Past research in which these mathematically supported statistical approaches have yielded erroneous conclusions—such as through underestimating outliers—only shows that this research is—as it should be—informative yet fallible. The flex-

---

[5] See Taleb, Nassim Nicholas, 2012, *Antifragile: Things that Gain from Disorder*, New York: Random House, pp. 339, 346, 347, and 358; see likewise Silver, Nate, 2012, *The signal and the noise: why so many predictions fail—but some don't*, New York: the Penguin Press, p. 183. The Economist article is "Unreliable research: Trouble at the lab, Scientists like to think of science as self-correcting. To an alarming degree, it is not," October 19, 2013.

[6] See Taleb, Nassim, 2012, Ibid., pp. 138, 288, 288, 349, 350, 416, and 417; for an account of the "Black Swan," see Taleb, N. N., 2007, *The Black Swan: The Impact of the Highly Improbable*, New York: Random House.

[7] See Freedman, David and Diana B. Pettitti, 2002, "Salt, Blood Pressure, and Public Policy," *International Journal of Epidemiology*, vol. 31 (2002), pp. 312–320.

ibility of mathematics in constructing distributions facilitates looking at issues from the perspective of multiple models.

The next chapter will provide needed balance to these perspectives that rightly stress the negatives of overly linear accounts of statistics as they are applied in medicine and elsewhere. However, one cannot have balance if one expects statistical findings to be incorrigible—in spite of future events and consequences that may not be very foreseeable or at least not foreseen.

6.4 Sample Linear Methods Devised by R. A. Fisher and Others to Bridge the Gap between Distributions, with Infinite Populations and Finite Data Samples

Of special interest throughout this document is the assumption of unique answers in much statistical practice. One may consider two general methods that have been developed in order to bridge the gap between distributions with infinite populations and finite data samples. One of these methods consists of the use of various types of significance tests. A second consists of methods that are used to "fit" models to the data and to account for uncertainties in these data. For considering these two types, we start with the statistical procedures developed by R. A. Fisher, whose outlook could be considered to be almost exclusively Gaussian and who insists on statistics that yield unique solutions. He proclaims:

> **Consistent** statistics ...all tend more and more nearly to give the correct values, as the sample is more and more increased; at any rate, if they tend to any fixed value it is not to an incorrect one.

That is, Fisher maintains that these methods yield convergence or single answers.[8]

Building on previous work by Karl Pearson and many others, R. A. Fisher calls "fiducial probability" methods for evaluating hypotheses based on only a finite number of samples. Originally the chi-squared test by Pearson had yielded one method for evaluating hypotheses. This test assumed that the observed frequencies tended toward the chi-squared distribution as the sample became larger. The table of the chi-squared test gives the area of the tail of a continuous curve.[9]

Fisher was concerned principally with three very common underlying distributions that apply to a great variety of the statistical subject matter: the binomial, the Poisson, and the normal (Gaussian). As with the chi-squared test, the goal of the use of these distributions was to supply a distribution, assumed to have an infinite population, to a hypothesis with a finite sample in order to find out the significance of this finite sample.

---

[8] See Fisher, Ronald Aylmer, 1944, *Statistical methods for research workers*, London: Oliver and Boyd Ltd., ninth edition, p. 11.

[9] See Fisher, Ronald Aylmer, pp. 10, 41, 92, and 93.

6.5 Selected Criticisms of Significance Tests Used in Standard Statistical Work

Fisher also discusses various fitting techniques such as the method of maximum likelihood—always an "efficient" statistics, of which for him the least-squares method is an example.[10] N. Silver, famous for his forecasting feats, has even criticized these "fitting" procedures developed by R. A. Fisher (and others) and followed by many as standard procedures in statistics. According to Silver, these "fitting" procedures presuppose an underlying distribution. Similarly, the Kolmogorov-Smirnov test at first assumed the Kolmogorov distribution, a quasi-Gaussian distribution, but now in application can have such an underlying model as the exponential distribution.[11]

The late Berkeley statistician Freedman goes further in stating that the notions of "almost surely" (widely used in books on extreme value statistics) and the limiting relative frequency (used in the frequency and Bayesian theories, respectively) "are features of your opinion not of any external reality. ("Almost surely" means with probability 1, and the probability in question is your prior.)"[12]

D. Kaye and D. Freedman ask the question consistent also with Chaps. 2, 3, 4, and 5, "would a pattern wash out if more data were collected?" With respect to any selected prior distribution used to develop "significance test," this question indicates that the pattern in the finite sample may with more data become (a) more like the underlying presupposed distribution or (b) less like the underlying presupposed distribution. Moreover, in their critique of "significance" tests, Kaye and Freedman maintain that "if results are significant at the 0.05 level, it is tempting to conclude that the null hypothesis has only a 5 % chance of being correct."[13]

In developing "tests" to indicate whether or not there is a given "fit" or that a hypothesis is acceptable, R. A. Fisher's restriction of his "significance tests" to three distributions is very biased. Bayesians generally have noted that throughout his work, R. A. Fisher postulates the application of distributions that for Bayesians should be regarded as priors. Bayesians apply this as well to Fisher's maximum-likelihood method.[14]

---

[10] See Fisher, Ronald Aylmer, pp. 9, 19, and 21.

[11] See Silver, Nate, 2012, op. cit., pp. 251ff.; see also Wikipedia, "Kolmogorov-Smirnov test," accessed February 27, 2013.

[12] See Freedman, David, 1995, "Some issues in the foundation of statistics," *Foundations of Science*, vol. 1, pp. 19–83, reprinted in *Some Issues in the Foundation of Statistics*, Kluwer, Dordrecht (1997), Bas C. van Frassen, ed., p. 11.

[13] See Kaye, David H. and David A. Freedman, eds., 2011, "Reference Guide on Statistics," pp. 211–301, in *Reference Manual on Scientific Evidence*, Washington, D. C., National Research Council, Committee on the Third Edition of the Reference Manual on Scientific Evidence & Federal Judicial Center, pp. 115, 124, and 125. This article contests the view in The Economist, 2013, op. cit., to the effect that a 5 % significance level has some probability meaning, such as that there is a 5 % chance that the hypothesis is incorrect.

[14] S. McGrayne, op. cit., cites various Bayesians on these issues on, for instance, pp. 53 and 132.

It is possible to construct a very large number of distributions beyond those 40 or so that are very familiar to those who study distributions. *This notion of constructing a very large number of distributions is discussed, for instance, in earlier papers by Diaconis and Freedman, and many others.*[15] The presence of many possible underlying or prior distributions suggests that there may be multiple ways of "fitting" and also that there may be many hypotheses that are about as valid as one another.

Following Lincoln Moses, Kaye and Freedman maintain

> A given data set can be viewed from more than one perspective, can be represented by a model in more than one way. Quite commonly, no unique model stands out as "true" or correct, just so strong a conclusion might require a depth of knowledge that is simply lacking. So it is not unusual for a given data set to be analyzed in several apparently reasonable ways....Desirable features of a model include (i) tractability (ii) parsimony, and (iii) realism.[16]

6.6 Further Discussions of "Fitting" Procedures

When do professionals working on such major projects develop different statistical approaches even when faced with the same or very similar data sets? Part of the answer to this question comes from discussions in Menke's approach to discrete inverse modeling. For matrices, if the number of model parameters M equals the number of data points N, then there is a unique solution. If M > N, then the solution set is undetermined, and there is a solution only if "prior information" supplements the N data points. If M < N, then the problem becomes one of "fitting" N data to the model with M parameters. Cases in which M > N can arise if there is an "overfitting" of the data, although overfitting can also occur if too many parameters are used to solve a problem and this results in various types of confusion and inefficiencies.[17]

For N. Silver, if too few parameters are used, there is underfitting: not capturing as much of the signal as you should. If too many parameters are used, the fit is too tight: noise in the data is fit rather than discovering the underlying structure of the data. The overfit model that fits each individual sequence lowers the variance. Thus, minimizing variance cannot be the sole criterion used in fitting.[18]

---

[15] Hill, Theodore P., and David E. R. Sitton, 2004, "Constructing Random Probability Distributions," *Abstract and Applied Analysis*, 453–468, Chuong, Nurenberg, and Tutschek, ed., World Scientific Press. In order to construct various alternative distributions, merely normalize the results of $Y= a0 + a1*x + a2*x**2 + \ldots anx**n$ for any n so that the sum is 1.0 for x in $U(0,1)$. One may use $ai \geq 0$ to simplify the process. As higher-order equations are used, chances are increased that extreme value distributions can be produced, with their attendant "stability" issues.

[16] See Kaye, David H. and David A. Freedman, 2011, Ibid., p. 120.

[17] See Menke, W., 1989, *Geophysical Data Analysis: Discrete Inverse Theory*, San Diego: Academic Press, pp. 39–49 and 89, 90.

[18] See Silver, Nate, 2012, op. cit., pp. 163 and 166–167.

As Silver discusses overfitting and underfitting, he further criticizes the development of fitting tests by R. A. Fisher. Not only does Fisher require underlying distributions, but Silver contends that as a frequency theorist Fisher emphasizes the objective purity of the experiment (in contrast to the theory-laden view of N. R. Hanson). In addition, tests of significance, often the binary null hypothesis tests, are too "clumsy" for gambling. One should also be reminded that the centile ranks assigned to null hypotheses have never been confirmed.[19]

For cases in which $M < N$, Menke distinguishes among L1, L2,..., Lm approaches to "fitting" data to models. For instance, L1 represents the use of deviations between a parametric form and the data to be fit. L2 represents the use of deviations squared between a parametric form and the data to be fit. For Menke, the typical or Gaussian approach uses L2 as a means of fitting data, typically to one that reduces the total sum of deviations squared, the least-squares method. For Gaussian distributions, this method turns out as well to be the maximum-likelihood method, the method that achieves the highest likelihood of a fit for all data.[20]

As Menke moves to a heavier-tailed distribution, the exponential distribution, he uses L1 instead of L2 norms to evaluate "fits" to data. Minimizing average deviations thus becomes the selected norm. The L1 norm weights "bad" data less than any Li norm in which $i > 1$. For Gaussian unstable distributions (the variance is infinite), the use of the L1 norm is also common. As may be expected from the mathematization of statistics and the ready availability of the Internet, a very large number of "fitting" programs are now available for a small number of extreme value mathematical forms.[21]

Mathematization has from nearly the outset been applied to extreme value distributions. Some have been defined as being "stable." To achieve this, Nolan extends this to the use of L0.8 and so on fitting techniques in cases in which, for instance, the Poisson slope is 0.9. This stratagem is required because L1.0 will not work if the slope of the Pareto distribution, for instance, is below 1.0. Still, Nolan's stratagem depends on knowing what this slope actually is—in the long run. Recall that only a finite number of samples are used to define any such stable distribution and further data may be expected to yield different results and in some cases extremely different results.[22]

---

[19] See Silver, Nate, 2012, Ibid., pp. 251–261; see also Hanson, Norwood Russell, 1965, *Patterns of Discovery: An Inquiry into the Conceptual Foundations of Science*, Cambridge: Cambridge University Press.

[20] See Menke, W., 1989, op. cit., pp. 27, 37, 80, and 83 and Neter, John et al., 1985, op. cit., p. 50.

[21] See Menke, W., 1989, pp. 133 and 141.

[22] See Nolan, John P., 2009, *Stable Distributions: Models for Heavy Tailed Data,* accessed on the Internet February 27, 2013, p. 15, and Wikipedia, "Stable Distribution," accessed February 27, 2013.

6.7 Summary of the Mathematization of Statistics

The mathematization of statistics attempts to bridge the gap between finite samples and the infinite populations assumed by probability and statistical theories. The mathematization of statistics provides a cornucopia of valuable tools that are often being augmented. Tools of interest in this chapter pertain to significance tests and fitting methods. These tests provide diagnostics for hypothesis testing and fitting of models. Yet, as N. Silver has asserted, these tests require underlying distributions. For R. A. Fisher, an early developer of these bridging methods, the chief underlying distributions are the binomial, Poisson, and normal (Gaussian). Gaussian methods placing high stress on variance and the size of the sample permeate Fisher's embryonic writings. These underlying distributions assumed by Fisher have been expanded so that—to repeat—the Kolmogorov-Smirnov test, once associated with the light-tailed Kolmogorov distribution, has been expanded to permit other underlying distributions such as the exponential distribution.

The theory of random probability distributions has been expanded so that indefinitely many distributions—far more than the 30, 40, or so that are familiar—can be constructed. This expansion of the theory of random probability distributions now permits investigators to postulate alternative underlying distributions with alternative solutions. The tests posited by R. A. Fisher and followers can be used as fallible diagnostics about whether or not the underlying distribution "washes out" with future data. The use of multiple underlying distributions belies R. A. Fisher's dictum that there is one correct solution.

The lack of unique outcomes can also be shown as one examines the flexibility that different investigators have in applying statistics. This is shown in inverse modeling. Also, similar problems of underfitting and overfitting apply in regression and other modeling.

Any hubris associated with existing tests and fitting procedures should be lessened through the recognition of the errors that have arisen from the standardized ways of using statistics. New cases or sequences of cases may always arise that modify previous results considerably. As with life-expectancy data, striking modifications may occur, for instance, when infant mortality rates are reduced considerably. Other extreme examples of these cases, as Taleb points out, can initially suggest extreme skepticism about the statistical use of data and the compilation of huge amounts of data.[23] More positively, this chapter stresses the multiple interpretations that can and should be derived from the use of statistics. The next chapter will evaluate major successes in the use of statistics and show different and more nuanced ways to interpret statistical findings.

[23] See Silver, Nate, 2012, Ibid., pp. 251–261; see also Hanson, Norwood Russell, 1965, *Patterns of Discovery: An Inquiry into the Conceptual Foundations of Science*, Cambridge: Cambridge University Press.

References

Fisher, R. A. (1944). *Statistical methods for research workers* (9th ed.). London: Oliver and Boyd Ltd.

Freedman, D. (1995). Some issues in the foundation of statistics. *Foundations of Science, 1*, 19–39.

Freedman, D., & Petitti, D. B. (2002). Salt, blood pressure, and public policy. *International Journal of Epidemiology, 31*, 311–315.

Hill, T. P., & Sitton, D. E. R. (2004). Constructing random probability distributions (with D. Sitton), *Abstract and Applied Analysis,* 453–468.

Kaye, D. H. & Freedman, D. A. (2011). "Reference Guide on Statistics," In *Reference manual on scientific evidence.* Washington, DC, National Research Council, Committee on the Third Edition of the Reference Manual on Scientific Evidence & Federal Judicial Center.

Keynes, J. M. (1921). *A treatise on probability*. London: MacMillan and Co.

Mandelbrot, B. B. (1983). *The fractal geometry of nature*. New York: W. H. Freeman and Company. (originally 1977)

Mandelbrot, B. B. (2012). *The fractalist: Memoir of a scientific maverick.* New York: Pantheon Books.

McGrayne, S. B. (2011). *The theory that would not die: How Bayes' rule cracked the enigma code, hunted down Russian submarines & emerged triumphant from two centuries of controversy.* New Haven: Yale University Press.

Menke, W. (1989). *Geophysical data analysis: Discrete inverse theory*. San Diego: Academic.

Neter, J., Wasserman, W., & Kutner, M. (1985). *Applied linear statistical models.* Homewood: Irwin.

Nolan, J. P. (2009). *Stable distributions: Models for heavy tailed data.* Accessed on the internet 27 Feb 2013.

Silver, N. (2012). *The signal and the noise: Why so many predictions fail—But some don't.* New York: The Penguin Press.

Taleb, N. N. (2007). *The black swan: The impact of the highly improbable*. New York: Random House.

Taleb, N. N. (2012). *Antifragile: Things that gain from disorder*. New York: Random House.

Wittgenstein, L. (1967). *Remarks on the foundations of mathematics*, Cambridge, MA: The M. I. T. Press, first published in 1956, edited by G. H. von Wright, R. Rhees, G. E. M. Anscombe, and translated by G. E. M. Anscombe.

Part IV
New Ways of Thinking

Chapter 7
Robust Simulation and Nonlinear Reasoning

And how does it come out that the proof compels *me? Well, in the fact that once I have got it I go ahead in such-and-such a way, and refuse any other path. All I should further say as a final argument against someone who did not want to go that way, would be "Why, don't you see…!"—and that is no* argument. *"But, if you are right, how does it come about that all men (or at any rate all normal men) accept these patterns as proofs of these propositions?"—It is true, there is great—and interesting—agreement here. (From p. 13e, Wittgenstein, Ludwig, 1967,* Remarks on the Foundations of Mathematics, *Cambridge, MA: the M. I. T. Press, first published in 1956, edited by G. H. von Wright, R. Rhees, and G. E. M. Anscombe and translated by G. E. M. Anscombe)*

Abstract In this chapter, robust simulation is introduced as a methodology that augments and supersedes previous approaches. Robust simulation is the representation of future risk through simulation of an ensemble of credible views that integrates valid scientific disagreement. From the standpoint of calculation, robust simulation supersedes the use of variance, confidence intervals, and other methods that have been used to develop uncertainty estimates. Robust simulation also augments the mathematical tools discussed in previous chapter through the presence of a community of investigators on major issues and the use of models derived from various disciplines. In competitive settings for addressing these socially important risk issues, alternative investigators may arrive at different risk estimates. The ensemble of these different estimates characterizes the current bounds of their uncertainty.

Robust simulation requires, moreover, intelligence applied to problems at hand and encourages competition among investigators. This competition will lead to an improvement in models used but not necessarily to convergence among any reasonable alternative approaches. Huge data samples may be helpful but are often not required to develop adequate statistical findings. These multiple interpretations show how alterative investigators can provide an ensemble of definite outcomes. Suggesting that there is always but one outcome is to endorse a false precision. The stochastic interpretations of complex phenomena clarify the range of current outcomes.

C.E. Taylor, *Robust Simulation for Mega-Risks*, DOI 10.1007/978-3-319-19413-4_7

7.1 Introduction to Robust Simulation

Robust simulation is the representation of future risk through simulation of an ensemble of credible views that integrates valid scientific disagreement.[1] Robust simulation augments the mathematical tools discussed in previous chapter through the presence of a community of investigators on major issues and the use of models derived from various disciplines. This community of investigators on major issues can be thought of as being both competitive and cooperative and enhance a view of science as defined by Imre Lakatos who states:

> *The history of science has been and should be a history of competing research programmes (or, if you wish, 'paradigms'), but it has not been and must not become a succession of periods of normal science: the sooner competition starts, the better for progress.*[2]

In other words, the presence of multiple approaches assists rather than hinders progress. The multiple approaches of merit are those developed by professionals and are heavily dependent on the experience of these professionals. These professionals may derive different approaches in parameters needed, applicable tests, pertinent data, data required, mathematical forms of models used, suitable decision models, and the science desired and needed to render the models credible. It is this community of investigators with a variety of approaches and goals from various disciplines who provide the credibility of robust simulation.

One may contend that robust simulation is an approach, a process, a strategy, or a more global outlook, but at the same time it can provide a "model," a method, or something confined even to a single investigative team. Whether performed by multiple investigative teams or by a single team, the result is a range of estimates (not confidence intervals in the traditional sense) that can be established in reference to a critical query.

Previous chapters have indicated some of whys and hows of robust simulation. This chapter begins with brief references to de facto illustrations of robust simulation, starting from illustrations in the social sciences and progressing to some more detailed quantitative examples. Afterward this chapter connects the topic of robust simulation to the topic of nonlinear reasoning—already introduced in various previous chapters—as in brief discussions of the development of Newtonian science and

---

[1] Lee, Yajie, Taylor, Craig E., Hu Zhenghui, Graf, William P., and Huyck, Charles K., 2014, "Using Robust Simulation to Characterize Uncertainties in Catastrophe Loss Assessments" from RAA Cat Modeling 2014, ImageCat, Inc.

[2] From p. 69, Lakatos, I., 1978, *The methodology of scientific research programmes*, London: Cambridge University Press. In contrast to theories of data mining, Lakatos has provocatively suggested: *as science grows, the power of empirical evidence diminishes* (p. 21). The presence of the ability to store (e.g., the "cloud") and process immense amounts of data, though, implies that one can, for instance, provide controls to telecommunications systems that lack the systemic weaknesses of typical hub-and-spoke systems. (See, for instance, Tang, Alex K. et al., 2014, "Lifeline System Interdependencies—Key for Resilience in Practice," *the Second International Conference on Vulnerability and Risk Analysis and Management (ICVRAM2014) and Sixth International Symposium on Uncertainty Modelling and Risk Analysis (ISUMA2014)*, Liverpool, July.)

the calculus. During the process of the testing and development of a fruitful idea, ranging from days to months to years to centuries and even millennia, many competing viewpoints have room for development and even the final success of many critical ideas does not discourage further qualifications and developments.

7.2 Robust Forecasting in the Social Sciences, Betting, and More Qualitative Studies

The notion of robust simulation begins with processes for which (1) simulation is extremely important to develop "statistical" or "risk-based" outcomes and (2) alternative and often very complex models are used to develop a set of reasonable outcomes. These procedures are very valuable when as at RAND large-scale computational resources are available yet models being used—in these cases largely economic and policy related—may vary significantly from investigator to investigator. An example of these uses of robust simulation is found in Lempert et al., which takes on 100-year economic forecasting.[3]

A second application of robust simulation derives from forecasting methods associated with betting. In betting (often referenced as a "test" of a statistical finding) requiring skill, some practitioners of course have better results than others. Nate Silver is one whose bets (as in the 2012 US Presidential election) have turned out to be notoriously and basically right.

Much of Silver's 2012 book describes a theory of forecasting. Unlike Taleb, Silver forwards a positive viewpoint on forecasting rather than extreme skepticism. Silver's theory has two major features consistent with this chapter[4]:

- First, statistics and forecasting should be regarded as being fallible activities. There are no perfect data or models. Instead, there are problems to be solved and questions to be addressed. Strong statistical inferences are backed up by theory or at least some deeper thinking about root causes.[5]
- Second, aggregate forecasts are "better." Some of these aggregate forecasts may be "better," but the overall aggregation of forecasts provides better forecasts overall. The individual forecasts made that are aggregated need to be in some sense "independent."[6]

---

[3] See Lempert, R. J., Popper, S. W., and Bankes, S. C., 2003, *Shaping the Next One Hundred Years: New Methods for Quantitative, Long-Term Policy Analysis*, Santa Monica, CA: RAND, and Lempert, R. J, Groves, D. G., Popper, S. W., and Bankes, S. C., 2006, "A General, Analytic Method for Generating Robust Strategies and Narrative Scenarios," *Management Science*, Vol. 52, No. 4, April, pp. 514–528.

[4] See Silver, Nate, 2012, *the signal and the noise: why so many predictions fail—but some don't*, New York: the Penguin Press.

[5] From p. 333, Silver, Nate, 2012, Ibid.

[6] From pp. 73, 335, Silver, Nate, 2012, Ibid.

Unfortunately, Silver's view of "robust" forecasting is aligned with a view of Bayesian statistics that assumes that individual forecasters will have their views "converge" as time goes forward. (See the previous discussions in Chap. 4.) This view of convergence is aligned with a linear view of scientific progress that assumes that we are converging to the truth as data are compiled. Bayes himself shared an eighteenth-century underlying deterministic world view. Silver's view also assumes that one can develop confidence intervals from the aggregate forecasts. This presumes that the law of total probability obtains, even though it is a separate and often daunting exercise to determine that one has assembled all candidate forecasts.[7]

The view developed in this chapter is that ensembles provide the risk solutions in robust simulation. Multiple perspectives do not need to be and probably should not be entirely independent—especially from the disciplines with which they are associated. How one weights these ensembles is an open question inasmuch as values close to the highest may be the best in some cases yet those values close to the lowest may be the best in other cases.

In addition, many examples come from a large number of studies concerned with the environment. Risk-based approaches on more qualitative topics, such as portions of geology and also water management areas, may likewise be familiar with dealing with diverse opinions and not expecting precise results. The view that multiple perspectives pertain to scientific activities is familiar in ecological studies. Stephen Toulmin has contrasted the modern view, from the seventeenth century through 1950 or after, according to which unique perspectives were solely required in critical disciplines. In accounting for some of the changes to this modern view, Toulmin cites ecological studies:

> As late as 1960, the word "ecosystem" had not yet won a place in the political vocabulary of industrial nations. John Muir and Aldo Leopold had crusaded for the environment, and for the threatened populations of endangered species. But Rachel Carson's book *Silent Spring* first spoke, in 1962, to the entire public audience—that is to an audience that was now ready to hear its message.[8]

Thus, robust simulation approaches may seem to be passé to many investigators in ecological studies and critical studies generally that are very qualitative.

7.3 Robust Simulation in More Quantitative Areas of Science and Engineering

Risk and decision models for evaluating complex systems require a great many submodels covering:

[7] From pp. 242, 410, Silver, Nate, 2012, Ibid.

[8] From p. 163, Toulmin, Stephen, 1992, *Cosmopolis: The Hidden Agenda of Modernity*, Chicago: University of Chicago Press. See also Carson, Rachel L., 1950, *The Sea Around Us*, New York: Oxford University Press, 1989 edition.

- Exposure data for sites, structures, system, people, animals, and habitats
- The peril and its frequency of occurrence (for different "severities")
- The transmission and distribution from the source(s) to site(s) resulting in "site intensities"
- The direct response of structures, people, and animals to the distributed intensities
- Any further perils that arise as a consequence of the initial peril
- The indirect response of systems given these direct responses
- The community and economic changes given the direct losses and the indirect response of systems
- Alternative ways to reduce direct and indirect damages, respectively
- Decision methods for evaluating these alternative ways to reduce damages

Each of these sub-model regions could be discussed extensively.

In general, the sub-models needed for a full-scale risk and decision analysis have always exhibited an unevenness in development. As discussed in many works by Graf et al., even sub-models with extensive use and refinement may be subject to nonlinearities and other issues that strongly suggest further needed work.[9] This unevenness gives rise not only to needed research but also to the presence of alternative models covering the same sub-model areas. For instance, different ways to gather exposure data exist especially as (1) more data are electronically maintained but still need to be curated and (2) much data may be proprietary.

For many of the sub-models, uncertainty evaluations (called here "endogenous uncertainties") are desirable. For instance, the Beta distribution may be used to evaluate "endogenous" uncertainties in the direct response of structures to intensities. Endogenous uncertainties can in theory be calculated in run time for the full evaluation of a risk and decision model. In this simulation process, these endogenous uncertainties "vanish" as more simulations are used (unless, as explained in Chap. 5, the strong law of large numbers does not prevail).

If these risk and decision evaluations were linear, and not systemic, then the weak sub-models would subvert the general evaluations. Since there are many sub-models, and as already indicated a great deal of flexibility in model development, there have been in many cases multiple models that can be used.

In contrast to the theory of statistics proposed by von Mises and studied in Chap. 3, there are exogenous probabilities required that do not necessarily follow a binomial model. In particular, frequencies of occurrence of the peril are exogenous estimates. These are required for estimating risk in a minimal sense. If, for instance, one investigator provides one "event set" for perils (e.g., one event set for hurricanes impacting a specific region), then to combine this with another "event set" or to vary the probabilities for each sample could create considerable confusion. Alternative models of the frequencies of the peril, then, are a major source of alternative outcomes.

So, (de facto) cases in which multiple (quantitative) models are common include:

- Catastrophe risk models for such perils as flood, earthquake, and hurricane[9,10]
- Global climate change models downscaled for different regions[11]
- Missile risk evaluations (and related asteroid risk evaluations)[12]

There are alternative decision theories that can be adapted to many situations, including those in which there is no simple "optimal" decision.[13]

Using earthquake risk evaluations alone, one can find in the literature an enormous number of alternative models. But a small number of these include:

- A variety of different inverse modeling approaches and outcomes on modeling earthquake events in great detail, such after they have occurred or in the laboratory[14]
- Very significant differences on modeling seismicity, including time between major events and general regional energy modeled[15]

---

[9] See Craig Taylor, Yajie Lee, William Graf, Zhenghui Hu, and Charles Huyck, 2010, "Robust Simulation and Cat Diagnostics for Treating Uncertainties in Catastrophe Risk Analysis," pp. 155–163 in The 1994 Northridge, California earthquake: Investigation of rupture velocity, risetime, and high-frequency radiation *of the International Symposium on Reliability Engineering and Risk Management*, ed. by Jie Li, Yan-Gang Zhao, Jianbing Chen, and Yongbo Peng, Shanghai, China, Tongji University Press; Craig Taylor, William Graf, Yajie Lee, Charles, Huyck, and Zhenghui Hu, 2011, "Sample Treatment of Uncertainties in Earthquake Portfolio Risk Analysis," pp. 246–251 in *Vulnerability, Uncertainty, and Risk: Analysis, Modeling, and Management*, Proceedings of the ICVRAM 2011 and ISUMA 2011 Conferences, edited by Bilbal M. Ayyub, Reston, VA: American Society of Civil Engineers; William Graf, Yajie lee, Charles Huyck, and Zhenghui Hu, 2012, "Propagation of Uncertainties through Robust Simulation and Future Research," *Fifth Asian-Pacific Symposium on Structural Reliability and its Applications* (*5APSSRA*), Phook, K. K., Beer, M., Quek, S. T., and Pang, S. D., editors, Singapore; and C. Taylor, R. Murnane, W. Graf, and Y. Lee, 2013, "Epistemic Uncertainty, Rival Models, and Closure," *Natural Hazards Review*, February, pp. 42–51, volume 14, number 1.

[10] See R. J. Murnane, C. E. Taylor, T. Jagger, and Z. Hu, 2011, "Robust simulation for sensitivity analysis of catastrophe risk losses," in *Applications of Statistics and Probability in Civil engineering*, ed. M. H. Faber, J. Koehler, and K. Nishijima, CRC Press, New York, PP. 875–877.

[11] Examples of ensemble climate change outcomes are made in many works including Alexander A. Golub, and Anil Markandya, 2008, *Modeling Environment—Improving Technological Innovations Under Uncertainty*, London: Routledge, Taylor & Francis Group.

[12] See Collins, Jon D. and Steven L. Carbon, 2010, "Launch Risk Acceptability Considering Uncertainty," *Proceedings of the 4th International Association for the Advancement of Space Safety* (*IAASS*) *Conference*, Huntsville, AL, 19–21 May.

[13] See Resnick, S., 2007, *Heavy-Tail Phenomena*, New York: Springer; Markowitz, H. M., 1959, *Portfolio Selection: Efficient Diversification of Investments*, Oxford: Basil Blackwell Ltd.; Levy, H., 2006, *Stochastic Dominance: Investment Decision Making Under Uncertainty*, 2nd edition, New York, NY: Springer; and Taylor, Craig, Glenn Rix, and Fang Liu, 2009, "Exploring Financial Decision-Making Approaches for Use in Earthquake Risk Decision Processes for Ports," *Journal of Infrastructure Systems*, Volume 15, Number 4, pp. 406–416, December 1, 2009.

[14] See Collins, Jon D. and Steven L. Carbon, 2010, "Launch Risk Acceptability Considering Uncertainty," *Proceedings of the 4th International Association for the Advancement of Space Safety* (*IAASS*) *Conference*, Huntsville, AL, 19–21 May.

[15] See Kagan, Yan Y., et al., 2007, "A Testable Five-Year Forecast of Moderate and Large Earthquakes in Southern California Based on Smoothed Seismicity," *Seismological Research Letters*, Vo. 78, No. 1, January/February 2007, pp. 94–98; National Research Council (NRC), (1997), *Review of Recommendations for Probabilistic Seismic Hazard Analysis: Guidance on*

- On modeling attenuation of seismic waves from source to site[16]
- On modeling vulnerability of structures to strong ground motions[17]
- On modeling higher-order economic losses resulting from catastrophic damage[18]

The number of types of sub-models needed and the number of pertinent references could be multiplied by several orders of magnitude. These sub-models could be increased a great deal for such infrastructure systems as those for electric power, culinary water, natural gas, highway, ports, air traffic, and telecommunications. These cases show the need to show that how, why, when, and where these multiple outcomes are possible.

7.4 Nonlinear Reasoning in Successful Cases Using Finite Samples and Experiments

Significance tests and fitting methods discussed in this and the previous chapter do not yet provide adequate credibility given the many alternatives to "standard" or Gaussian statistical procedures. The addition of professional groups and underlying

Uncertainty and use of Experts, Washington, D. C.: National Academy Press; Perkins, D., 2002, "Uncertainty in Probabilistic Seismic Hazard Analysis," pp. 19–60 in *acceptable Risk Processes: Lifelines and Natural Hazards*, edited by Taylor, C. and VanMarcke, E., Reston, VA: American Society of Civil Engineers; Petersen, Mark D., Tianqing Cao, Kenneth W. Campbell, and Arthur D. Frankel, 2007, "Time-independent and Time-dependent Seismic Hazard Assessment for the State of California: Uniform California Earthquake Rupture Forecast Model 1.0, *Seismological Research Letters*, Vol. 78, No. 1, January/February 2007, pp. 99–109; and Petersen, Mark D., Arthur D. Frankel, Stephen C. Harmsen, Charles S. Mueller, Kathleen M. Huller, Russell L. Wheeler, Robert L. Wesson, Yuehuan Zeng, Oliver S. Boyd, David M. Perkins, Nicolas Luco, Edward H. Field, Chris J. Wills, and Kenneth S. Rukstales, 2008, *Documentation for the 2008 Update of the United States National Seismic Hazard Maps*, US Department of the Interior, US Geological Survey, Open-File Report 2008–1128.

[16] See Bradley, Brendon A., 2009, "Seismic Hazard Epistemic Uncertainty in the San Francisco Bay Area and Its Role in Performance-Based Assessment," *Earthquake Spectra*, Vol. 25, No. 4, pp. 733–754, November; Stewart, Jonathan P., Ralph J. Archuleta, Maurice S. Power, 2008, "Special Issue on the Next Generation Attenuation Project," *Earthquake Spectra*, Vol. 24, No. 1, February; Strasser, F. O., Abrahamson, N. A., and Bommer, J. J., 2009, "Sigma: Issues, Insights, and Challenges," *Seismological Research Letters*, Volume 80, Number 1, pp. 40–56, January/February; Trifunac, M. D., 1997, "Stresses and intermediate frequencies of strong earthquake accelerations," Geofizika, vol. 14, pp. 1–27.

[17] See Cho, S., et al., 2006, "Calibration of Default Bridge-Damage Model," Appendix K in *REDARS 2 Methodology and Software forSeismic Risk Analysis of Highway Systems*, by Werner, S. D., et al., Buffalo, N.Y.: MCEER under FHWA Contract Number DTFH61-98-C-0094; Graf, William P., and Lee, Yajie, 2009, "Code-Oriented Damage Assessment for Buildings," *Earthquake Spectra*, Volume 25, No. 1, February; and Wesson, Robert L., et al., M. D., 2004. Losses to single-family housing from ground motions in the 1994 Northridge, California, earthquake, *Earthquake Spectra* **20,** No. 3, 1021–1045.

[18] See Rose, A. and S. Liao. 2005. "Modeling Regional Economic Resilience to Disasters: A Computable General Equilibrium Analysis of Water Service Disruptions," *Journal of Regional Science* 45(1): 75–112.

agreements again can be subjected to questioning even though these additions provide persuasive reasons for granting robust simulation outcomes credibility, at least within the state-of-the-practice or state of the art. The issue of finite samples still poses questions about robust simulation results.

A different approach to these issues stems from a discussion of successes in the application of statistics conjoined with qualitative considerations. This discussion contrasts to discussions of failures in Chap. 6. Successful experiments also provide insights into the role statistics can play in advancing knowledge. What follows, based on work by D. Freedman on statistics and R. Harre on experiments, provides but a small portion of the vast literature on these topics and topics related to the advances in knowledge generally. However, characteristic of these examples is the nonlinear character of overall processes of reasoning often initially associated with specific individuals but in all cases associated with eventual great interest scientific and/or practical by many people. These cases further illustrate how in nonlinear reasoning one does not begin with a solid initial truth whose consequences are known fully to everyone. Instead, the early version of a successful theory—if evaluated solely at that moment in terms of its "truth"—will typically be at best a fruitful direction but "false" if taken as a final proposition.

Freedman[19] uses the following celebrated case studies to illustrate the role that finite statistics and qualitative consideration play in addressing very significant medical issues (with the successes initially defined in terms of one investigator in spite of the long history before and after each success). Note that Freedman calls the cases those involving causal process.

The following nine case studies are in sharp contrast to pure data-mining activities.

7.4.1 Case Study: Edwin Jenner's Discovery that Injection of "Cowpox" Could Eliminate or Greatly Diminish Smallpox[20]

Smallpox elimination is often considered an unmitigated success, but it was actually a very complex history. Smallpox had arisen maybe in 10000 B.C. and had caused a great many epidemics over time. A method called variolation had been discovered that led to a reduction in smallpox deaths by a factor of about 10. This method had been used by the British and later by George Washington and his forces during the

[19] See Freedman, David A., d.u., "On types of scientific inquiry: nine success stories in medical research," in *Oxford Handbook of Political Methodology*, pp. 300–318, Janet M. Box-Steffensmeier, Henry E. Brady, and David Collier, editors.

[20] References for this discussion come from Wikipedia, "Smallpox," accessed July 13, 2013; see Wikipedia, "History of smallpox," accessed July 13, 2013; Wikipedia, "Edward Jenner," accessed December 1, 2013; and Riedel, Stefan, 2005, "Edward Jenner and the history of smallpox and vaccination," *BUMC Proceedings* 18:21–25.

revolutionary war. Before Jenner's sustained research efforts, cowpox was used by English doctors in regions with dairy farms and there were many tales that dairy-maids were protected from smallpox as a result of having suffered cowpox. Benjamin Jesty was among those who had discovered the efficacy of cowpox in eliminating or greatly diminishing smallpox. Against this background, Edward Jenner undertook case studies of the use of cowpox, and his first 40 of so case studies had proven to be successful: the "inoculation" of patients by cowpox had eliminated smallpox. Later studies led to at least one case in which the inoculation had only reduced the effects of smallpox. By 1800, Jenner was convinced that inoculation was effective in virtually eliminating deaths by smallpox. Owing to the slow testing needed, the British government eventually proscribed the use of "variolation" in lieu of the use of "cowpox" for reducing and eliminating smallpox. Over time, Jenner's procedures were tested on very large populations and their improvement and incorporation in more recent theories of the role of germs in disease, the discovery and study of viruses, and the developments of modern immunology. In 1958, smallpox still thrived in 63 countries. As a result of a comprehensive campaign to eradicate smallpox, by 1977 the World Health Organization had claimed that smallpox was eliminated. So, Jenner's finding conjoined with healthcare planning and execution has led to date to an unmitigated success, an extremely large population no longer suffering from smallpox. However, this may be the only case in which a human infectious disease has been completely eradicated.

Conclusion/Lesson Jenner's discoveries thus comprise a small portion of a much longer narrative on the use of cowpox to eliminate smallpox. Jenner's discovery ultimately competed very successfully in contrast to its competitor, variolation. Variolation remained a competitive alternative for many years. Large amounts of data have accumulated to indicate that smallpox has been eliminated. This case is as close as any of the cases discussed by Freedman involving indefinite progression or lawlike outcome from the work of Jenner and his many successors dealing with a problem of major social importance.

7.4.2 Case Study: Ignaz Semmelweis's Discovery that One Could Reduce Infectious Diseases Through Eliminating Cadaveric Particle Cleansing[21]

Serving as a professor's assistant in a clinic where both autopsies were performed and births occurred, Semmelweis noted that in one ward the fatality rate for women bearing children was much higher than in another. In effect women were dying at

[21] References for this discussion come from Wikipedia, "Semmelwise reflex," accessed December 2, 2013; Wikipedia, "Ignaz semmelwise," accessed December 2, 2013; and Robinson, Victor, 1912, "Pathfinds in Medicine: Semmelwise, the Obstetrician," *Medical Review of Reviews*, **18**, 232–245.

high rates for childbirth fever, called then "puerperal fever" and now "sepsis." Over time, he tried various prior theories that did not seem to apply to the discrepancy in fatality rates between the two wards: milk, local suppression, a gastric-bilious disturbance, peritonitis, being unmarried, and miasma (or "ill winds" from poor air ventilation). The disease was deemed to arise from poisonous gases from swamps, garbage pits, open graves, and rotting organic matter. Semmelweis noted that in one room students performed the birthing assistance, whereas in the other room midwives performed this assistance. Maintaining careful statistics, Semmelweis noted that women giving births out of doors had lower fatality rates. Over time, he decided to try disinfecting hands with a solution of chlorinated lime. Results in the ward with high fatality rates supported this hypothesis. He hypothesized that the cause of the high fatality rates was the absorption of putrid matter from a living organism or cadaver, producing a pyemic blood dissolution. The bacteria in the group *Streptococcus pyogenes* that caused the infection were identified only afterward. The theory of the role of germs in causing disease had not yet been accepted and studies of viruses and immunology did not at the time provide a theoretical underpinning of Semmelweis's results. As a consequence, in Europe his results were savagely opposed and this opposition has led to the "Semmelweis reflex." In England, a theory of "miasma," a theory of contagion, was at that time thought to have the same result as Semmelweis's theory. Thus, it was not until later that Semmelweis's antiseptic approach was accepted.

Conclusion/Lesson Once again Freedman is discussing an issue of major social importance. Again, the theoretical underpinnings of the views discovered by Semmelweis were not known until work by many in later investigations. Once again, there is opposition. Semmelweis's findings indicate how successful inquiry in statistics can be extremely nonlinear. Competitive hypotheses and theories persisted until the germ theory overcame its own objections and became a theory explaining Semmelweis's findings.

7.4.3 Case Study: John Snow's Discovery that an Infectious Disease Can Be Prevented by Cleaning Up the Water Supply System[22]

Making discoveries from the cholera epidemic that began in 1849, John Snow too found his discoveries to be opposed by supporters of the miasma theory. As with Semmelweis's theories and Jenner's discoveries, Snow was opposed in his lifetime but later was vindicated by history.

[22] References for this discussion come from Wikipedia, "John Snow (physician)," accessed December 3, 2013; Vachon, David, 2005, "Doctor John Snow Blames Water Pollution for Cholera Epidemic," *Old News* 16 (8), 8–10, May and June; UCLA [1], "Competing Theories of Cholera,"

Starting from detailed studies of this epidemic, Snow discounted the miasma theory as an explanation: deaths should have developed for miners, those in a workhouse, and those working in a local brewery. Instead, very detailed studies led Snow to find that of the first 83 deaths, 73 had died near a water pump that drew water from the Thames. Starting with these first findings, Snow's discoveries were very nonlinear. The initial evaluation of the pump indicated that the sewage line was well below the pump and the pump lining was intact. Water samples from the pump were microscopically compared with samples from elsewhere, yet no differences were disclosed.

Later data indicated 197 deaths, all within a 3 min walk of the pump. In 1854 at the outset of the outbreak, an infant's diapers had been dumped into a drain 32 in. from the pump well, and the infant had died from diarrhea. Changing the pump handle had apparently removed the local cholera problem. However, it appears that the real cause was the change of an intake pipe at the water company serving the pump. And it is only after Snow died that the Italian Fillipo Pacini had made histological examinations of intestinal mucosa and had discovered a bacillus that had caused cholera. Thus, later historical developments in histological data and germ theory had obviated many of the objections to Snow's statistical interpretations.

Conclusion/Lesson Again, these findings comprise a major historical narrative that lives well beyond the inchoate and opposed findings of the investigator, here John Snow. Again, the theoretical underpinnings of his theory were not yet well developed. Data mining based on erroneous theoretical underpinnings can be very flawed.

7.4.4 Case Study: Christiaan Eijkman's Discovery that Diet Deficiencies (Thiamine) Resulted in Beriberi[23]

Among his many researches, Eijkman's most famous work came as a result of his studying beriberi, "a disease of the peripheral nerves" in Indonesia/Dutch East Indies. Alternative hypothesis for the cause of beriberi had been blood contamination, respiratory metabolism, perspiration, or seasonal or temperature variation. Several of Eijkman's experiments with rabbits, monkeys, and chickens provided no insight into causes of beriberi.

By accident, Eijkman had discovered that some chickens used in his laboratory had symptoms of beriberi when their feed had been leftover rice from military rations but had no such symptoms when they are fed military rice. The causative

www.ucla.edu/snow/fatherofepidemiology.hyml, accessed December 3, 2013; and UCLA [2], "Who first discovered cholera?," www.ucla.edu/snow/fatherofepidemiology.hyml

[23] References for this discussion come from Wikipedia, "Christiaan Eijkman," accessed December 4, 2013; Wikipedia, "Beriberi," accessed December 4, 2013; BetterMedicine, "Beriberi," http://www.localhealth.com/article/beriberi; and Medicine Plus, "Beriberi," http://www.nih.gov//medcineplus/article/000339.htm

feed was polished rice—rice with its husk removed in order to extend its lifespan, as opposed to the unpolished rice. Eventually it was discovered that the missing compound was vitamin B1 or thiamine. Later investigations have found that alcohol can inhibit the body's ability to absorb thiamine. Wet beriberi is now described as beriberi affecting the cardiovascular system, in contrast to the dry beriberi that affects the nervous system. Infantile beriberi can occur when the nursing mother is lacking in thiamine. Genetic beriberi is a rare condition in which people lose the ability to absorb thiamine from foods.

Conclusion/Lesson The nonlinearity of Eijkman's discovery and subsequent developments is shown in how the terms become modified in the process of discovery. The initial discovery confounded two terms: the "beriberi" that he was addressing was the "dry beriberi" and not "wet beriberi." The element that caused the dry beriberi was named "thiamine." Multiple sources of the dietary deficiency came out after Eijkman's work. As with other cases cited, the success is derived for his dogged inquiry to solve a problem of major importance, a dogged inquiry that led to findings that could be used by others as a viable starting point.

7.4.5 *Case Study: Joseph Goldberger's Discovery that Diet Deficiencies (Niacin, Vitamin B3) Led to Pellagra*[24]

Pellagra was identified in among Spanish peasants in Spain. It was though conclusively distinguished from leprosy in 1907. Between 1907 and 1940, approximately 100,000 people died chiefly in southern US states. In 1914, Goldberger was asked by the US surgeon general to investigate pellagra, an endemic disease in the southern USA. Previously held opinions stressed how pellagra was regarded as an infectious disease. However, Goldberger found that germs did not explain the disease. In mental hospitals and orphanages, inmates and orphans contracted the disease, but the staff never did. Goldberger than developed an experiment with a sample of two orphans and inmates of a mental asylum. Those fed a diet of fresh meat, milk, and vegetables did not contract pellagra, whereas those fed a corn-based diet did contract the disease. Goldberger surmised that diet not germs caused the disease. Then Goldberger developed a small-sample experiment on 11 healthy volunteer prisoners, who were fed a corn-based diet. Six of these contracted pellagra after 5 months. When fed a normal diet, the pellagra vanished. Still unconvinced, Goldberger, his wife, and an assistant experimented on themselves to derive the same conclusion. Enraged southerners including a South Carolina congressman opposed Goldberger's

[24] References for this discussion come from Kraut, Alan M., "Dr. Joseph Goldberger & the War on Pellegra," http://history.nih.gov/exhibits/goldberger/index.html; Diet.com," Goldberger, Joseph," http://www.diet.com/g/goldberger.joseph; Wikipedia, "Joseph Goldberger," accessed December 5, 2013; and Wikipedia, "Pellegra," accessed December 5, 2013.

conclusions. Before he died in 1929, Goldberger thought that vitamin B was the deficient element in the diet.

After Goldberger died, Conrad Elvehjem discovered the more specific cause was the dietary lack of the B vitamin niacin along with reduced levels of the essential amino acid tryptophan. Further discoveries included possible excessive intake of leucine, possible alterations in protein metabolism in disorders of such as carcinoid syndrome, and a deficiency in the amino acid lysine that leads to a deficiency in niacin. North American Indians had "nixtamalized" maize and so had avoided the niacin deficiency. Nixtamalization is the treatment of the grain with lime and an alkali and has been shown to make niacin available in corn. Pellagra is common in Africa, Indonesia, North Korea, and China, and the majority of patients are poor, homeless, alcohol dependent, or psychiatric patients who refuse food. This century there have been outbreaks in Angola, Zimbabwe, and Nepal.

Conclusion/Lesson As with the other cases, the discovery that diet in contrast to germs was responsible for pellagra is part of a longer and ongoing historical narrative that fills in many of the details about causes of pellagra and its sources related to niacin deficiency in diet. Goldberger's discoveries comprise a major stage in this longer narrative, a major stage that had it not led to further inquiries could have been very misleading.

7.4.6 *Case Study: Frederick McKay's Discovery that Led to the Use of Fluoride in Water Systems*[25]

In the nineteenth century, a German, Carl Erhardt, had recommended potassium fluoride supplements to reduce tooth decay and the British James Crichton-Borwne had proposed the reintroduction of fluoride into the diet for similar reasons. From 1901 to 1933, Frederick McKay undertook studies of the brown-stained teeth (fluorosis) in such locations as Colorado Springs, CO, Oakley, ID, and Bauxite, AK. In the course of these studies, by 1917 McKay and colleagues accidentally found that lower rates of tooth decay were concomitant with brown-stained teeth. McKay had tested water samples but had found no difference in the water in these communities. In 1931, H.V. Churchill had used a more refined procedure, photospectographic

[25] References for this discussion come from Wikipedia, "History of water fluoridation," accessed December 6, 2013; Wikipedia, "Water fluoridation," accessed December 6, 2013; The Savvy Sister, "Dr. McKay fluoride," accessed December 6, 2013; William James Maloney and Maura Maloney, 2009, "Dr. Frederick McKay: Father of Communal Fluoridation," *Journal of the Massachusetts Dental Society*, vol. 58/no 1., Spring; Gower, Timothy, 2002, "A History of Fluoride," in *Prevention*, http://www.prevention.com; Medical Discovery, "Trustworthy Endodontist," http://www.discoveriesinmedicine.com/Enz-Ho/Fluoride-Treatment-Dental.html, accessed December 6, 2013; and NIH, National Institute of Dental and craniofacial Research, "The Story of fluoridation," http://nider.nih.gov/OralHealth/Topics/Fluoride/TheStoryofFluoridation.htm

analysis, and had found fluoride in the water samples. H. Trendley Dean at NIH began epidemiological studies of fluorosis. From 1933 to 1945, a Danish investigator Kaj Roholm that detailed some negative features of fluoride: bone disease, skin lesions, and mortality. But 1939, Gerald J. Cox had performed a study of the use of fluoride in rats and had concluded that their teeth were healthier. By 1942 an NIH study of 30,000 schoolchildren indicated that 1 ppm of fluoride in water was enough to minimize tooth decay yet not cause fluorosis. In 1945, Grand Rapids MI undertook a fluoridation program for its water system, and by 1956 a case study of 6000 schoolchildren was available. This epidemiological study indicated the water fluoridation reduced tooth decay by 2/3rds, a ratio higher than the 10–25 % reduction now estimated, with a range of −5 % to 64 % depending on circumstances.

Many scientific and social activities have been undertaken since these early studies. Now, for instance, it is known that there are many sources of fluoride along with fluoridation of water: natural water fluoridation in some cases, toothpaste, some air pollution, tea leaves, barley, and so on. In Sichuan, China, for instance, food is the main source of fluoride. Tolerable levels have been variously stated, possible at 0.01 mg/day for infants 6 months or less up to 0.1 mg/day for those 19 and above. Detailed chemical analysis of cavities has described how fluoride does not prevent but instead controls the rate at which cavities develop. Finland and Germany have stopped water fluoridation without increasing tooth decay rates. Yet, in many circumstances, water fluoridation is still regarded as being scientifically a sound method for reducing tooth decay.

Conclusion/Lesson Frederick McKay's discoveries comprise part of a long historical narrative that continues to have some degree of controversy long after his death. The many sources of fluoride along with issues pertaining to its safe limits have been discussed and continue to be discussed even today.

7.4.7 Case Study: Alexander Fleming's Discovery of the Mold Penicillin[26]

Before Fleming discovered a penicillin mold in 1928, many ancient cultures including Greeks and Chinese had used molds and other plants to treat infection and Serbs and Greeks used moldy bread as a traditional treatment for wounds and infections. In 1875, John Tyndall had described the antibiotic effects of *Penicillium* and in 1925, D. A. Gratia had done the same. Others having initial insights included Joaquim Caminhoa in Brazil, Vincenzo Tierio in Naples, Clodomioro Picado Twight in Costa Rica, and Ernest Duchesne.

---

[26] References for this discussion come from Wikipedia, "Alexander Fleming," accessed December 7, 2013; Bio, "Alexander Fleming biography, synopsis," http://www.biography.com/print/profile/alexander-fleming-92968941; New World Encyclopedia, "Alexander Fleming," http://www.newworldencyclopedia/org/entry/Alexander_Flemingl; and Wikipedia, "Penicillin," accessed December 7, 2013.

Fleming himself had found the antiseptics used in World War I to be ineffective because they dealt only with the surface infections. By accident after a vacation, Fleming noticed that one culture in his lab was contaminated with a fungus and that the colonies of *staphylococci* that had immediately surrounded it had been destroyed, unlike what happened in colonies farther away. The rare variant may have drifted from one floor below. He thought that he had discovered an enzyme but later called it a mold juice, named penicillin. Fleming was the first to isolate *Penicillium* and take it seriously. It turned out to be the first successful antibiotic.

But for some time he regarded its action to be slow and also short lasting and so ineffective for killing bacteria effectively. His clinical tests were inconclusive. Some later tests were more promising, but by 1939 the testing was taken up at Oxford by Howard Florey and Ernest Chaine, who stressed researching, isolating, and mass producing it.

Being regarded by some as one of the major medical discoveries of the last millennium, penicillin has been used for previously such serious diseases as syphilis and infections caused by staphylococci and streptococci. Millions of people may have been saved. Adverse effects have since been identified as has its correct structure and also its chemical mechanism of action. Numerous practitioners have played major roles in its development and successes. Yet, many types of bacteria have become resistant.

Conclusion/Lesson The historical narrative associated with Fleming's discovery contains an even more interesting consequence: dealing with bacteria reduces the changes that the penicillin vaccination will be sound indefinitely. The statistics and findings provided at one stage may prove to be very misleading as the bacteria produce defensive mechanisms for survival.

7.4.8 Case Study: Norman Gregg's Discovery that Exposure to German Measles in Early Pregnancy Led to More Infants with Cataracts and Heart Defects[27]

German or 3-day measles (rubella) had been described in the eighteenth century by Frederick Hoffman and considered a disease distinct from both measles and scarlet fever by George de Maton in the early nineteenth century. In the 1940s there was a sever outbreak of rubella in Australia. Norman Gregg began working on the issue of

[27] References for this discussion come from Wikipedia, "Norman Gregg," accessed December 8, 2013; Wikipedia, "Rubella," accessed December 8, 2013; Forrest, Jill M., Fiona M. Turnbull, Gary M. Sholler, Richard E. Hawker, Frank J. Martin, Margaret A. Burgess, and Trevor T. Doran, 2002, "Gregg's congenital rubella patients 760 years later," *Med. J. Aust.*, 177(11), 664–667; Dunn, P. M., 2007, "Perinatal lessons from the past: Sir Norman Gregg, ChM, MC, of Sidney (1892–1966) and rubella embryopathy," *Arch dis child Fetal Neonatal Ed*, 092: F513-514; and Louisiana Office of Public Health, Infectious Disease Epidemiology Section, 2012, "Rubella," July 2, 2012, www.infectiousdisease.dhh.louisiana.gov

problems in birth defects; the widely held assumption was that all such problems were genetic, not environmental. Gregg studied 78 children with congenital cataracts and found that 68 had been exposed to rubella in utero. Gregg and others continued these investigations and found that congenital defects included mental retardation, thrombocytopenic purpura, hepatitis, bone lesions, and meningoencephalitis were manifestations of the disease.

Later investigations have indicated that mothers infected with rubella within the first 20 weeks of pregnancy are especially susceptible to having infants die or have congenital problems. These include cardiac, cerebral, ophthalmic, and auditory defects and other possible defects. By 1961 investigators had isolated the virus RNA, toga virus genus *Rubivirus*, which is capable of crossing the placenta and infecting the fetus where it sops cells from developing or destroys them. In the 1964–1965 US epidemic of rubella, there were no fewer than 20,000 cases of congenital cataract. By 1969 a live attenuated *Rubella* virus vaccine had been developed. Since 1983 fewer than 1000 cases per year were reported in the USA. Later studies at the start of this century detailed the results of congenital rubella on those who had been studied by Gregg 60 years earlier. Ten had died and 32 were among those who volunteered for the study. Diabetes, thyroid disorders, early menopause, and osteoporosis had increased compared with the Australian population. Thus, Gregg's discovery along with later methods of prevention have greatly reduced the incidents of rubella and its adverse consequences.

Conclusion/Lesson Here again, the historical narrative of the discovery implies a change in what can count as a "cause" in congenital defects. And again, the medical profession has greatly advanced on the work of Gregg and his colleagues.

7.4.9 *Case Study Arthur Herbst's Discovery that an Artificial Hormone Led to Adenocarcinoma in Adolescent Girls*

In the late 1930s and early 1940s, obstetricians treated difficult pregnancies sometimes with a new wonder drug—diethylstilbestrol (DES), marketed by many companies. This is a nonsteroidal synthetic estrogen. The FDA had approved DES in 1940. When DES was suspected of causing cancer in the vagina, this was thought to be rare, and only for women over 50. Instead, from 1966 to 1969, seven girls ranging from 15 to 22 years of age from New England were found to have clear cell adenocarcinoma (CCA), and in 1969 another case arose for a 20-year-old patient in another Boston hospital. The clustering of cases led to considerable data gathering in order to identify why this clustering had occurred. Seven of the eight mothers volunteered that they had had stilbestrol in the first trimester of pregnancy; the other mother had been delivered in a private setting. Between 1946 and 1951 about 14,500 births were evaluated in one of the hospitals which had 675 cases in which stilbestrol was prescribed. Thus, a low ratio of women bearing children during this period had been treated with stilbestrol.

In April 22, 1971, Arthur Herbst and colleagues had written an article about how DES in mothers led to this carcinoma in their daughters. In 1971, the FDA rescinded its approval of DES. This discovery pushed medical research in the direction of studying prenatal environmental disruptor exposures to what had happened in the womb. For instance, vaginal cytology does not diagnose this carcinoma. Instead vaginal examination is required. Herbst's finding has led to many animal experiments dealing with the mechanisms of transplacental carcinogenesis and effects of exogenous hormones on the developing embryo. Third-generation effects have been studied, along with ways of educating the public and managing cases. Lawsuits against drug companies originally marketing DES have not proven to be effective.

Conclusion/Lesson This study begins with the approval of a drug that proves later to be harmful, an example of iatrogenics. It is not the "environment" or "genetics" that proves to be the source of vaginal cancer in young women but instead a drug used to assist their mothers in childbirth. As a consequence of Herbst's discoveries, many further findings have covered not only how other populations can have these cancers but also how to manage and treat those impacted by related health problems.

7.5 Lessons Learned from Complex Problem-Solving

In examining such cases of problem-solving, Freedman draws the following conclusions:

- The power of quantitative methods and good research design are important.
- Substantial progress derives from informal reasoning and qualitative insights.
- Recognizing anomalies is important.
- Investigators should have the ability to capitalize on accidents.
- Progress may require refuting conventional ideas if they are wrong and developing new ideas that are better and testing both new and old ideas. For instance, theories underpinning Semmelweis's discoveries had not yet validated his approach in spite of the data he had collected to establish his view.
- Given the nonlinearity of findings, the overly standardized use of statistics—with the appearance of rigor—can have negative impacts, such as missing insights developed through the use of finite numbers of carefully considered cases and failing to recognize determinants that have been ruled out in current prevailing models.

Amplifying Freedman's remarks, these cases of the successful use of finite statistics along with qualitative considerations show how:

- Sound but unfinished statistical and qualitative reasoning can become part of a longer major nonlinear historical narrative that responds to many issues in the understanding, clarification, and treatment of major health problems.

- During this longer and nonlinear historical narrative, there is opposition and at a minimum competition to the original insights mentioned.
- Thus, later developments become extremely important in these historical narratives; the findings do not stand by themselves, but their value or truth lies in their being taken up and augmented through further inquiry and practice.

In these case studies, data are used for further confirmation of findings derived through the application of problem-solving techniques (the use of intelligence, both qualitative and quantitative). But as the case of Semmelweis shows, data alone do not fully confirm these derivations. As Taleb has said,

> [p]eople are under the illusion that "science" means more data.... The more data...the more false information...The more data...the more correlations.[28]

At the same time, the investigators referenced above typically require further data to support and clarify their findings. Still, with only a finite amount of data available, there is always the strong possibility that the general discoveries will need to be qualified for some people or in some contexts.

Case studies of experiments can bring out similar findings. In Harre's account of 20 very famous experiments, made by investigators Aristotle and Theodoric of Freiburg to Pasteur and Faraday, we see the importance of treating experiments as steps in larger scientific programs. These experiments tend to add clarity to "vaguely delineating subject matter" of great importance in research. The qualitative nature of findings is supplemented by series of experiments that are not "isolated events that stand by themselves."[29]

7.6 Summary of Robust Simulation and Nonlinear Reasoning

Flexible mathematical tools are equipped to evaluate the acceptability of hypotheses and fits of data to diverse distributions. These tools, along with huge advances in information technology (IT), help greatly when dealing with multiple models, methods, approaches, processes, and strategies. From the standpoint of calculation, robust simulation supersedes the use of variance, confidence intervals, and other methods that have been used to develop uncertainty estimates. Confidence intervals and estimates of variance can vanish for Gaussian distributions and can wobble for extreme value distributions.

Robust simulation requires, moreover, intelligence applied to problems at hand. This intelligence heavily depends on piggybacking off a legacy of advances in diverse disciplines. A community of investigators is required along with competi-

[28] From pp. 128, 416, and 417 in Taleb, Nassim, 2012, op. cit.; similar remarks are found on p. 13, Freedman, op cit.

[29] From pp. 4, 5, Harre, Rom, 1983, *Great Scientific Experiments: Twenty Experiments that Changed our View of the World*, Oxford: Oxford University Press, pp. 4, 5.

tion among these investigators on matters of major social and economic importance.

This competition will lead to an improvement in models used, but not necessarily to convergence among any reasonable alternative approaches. Case studies in successful applications of statistics relative to field and/or experimental data can clarify that huge data samples may be helpful but are often not required to develop adequate statistical findings. The long historical narrative for major discoveries and their adumbration provides an explanation as to how competition is very common and needed during this long narrative. In most of the cases cited, this narrative is still ongoing. One expects far less than the full truth to be the starting point.

For instance, David Freedman gives a number of illustrations as to how finite and careful sampling can *play a key role* in developing broad solutions to critical problems—especially when this sampling is combined with different applications of intelligence that considers anomalies, modification of traditional views, serendipitous results, and even the application of a wide range of tools to solve problems. Success in these cases is defined in terms of achieving goals for a significant population, but it would be premature that success in these cases goes on forever.

In addition, R. Harre provides a similar account of great experiments. These cases arise in which there are major problems that need to be addressed, and the issue is not merely reaching the "truth" but also providing a broader solution to these problems. Samples are used to confirm and on occasion to discover the solution, but the active mind is required to deal with an array of modifications of previous results that have not provided the needed solution. Anomalies may prove to be important. Serendipitous results may be important. Modifying conventional views may be needed. Viewing cases of successes in statistics and experimentation provides insight into the limitations of extreme skepticism, which, while it may be true in the absolute, does not assist when one is trying to provide broader solutions to currently unsolved problems. Solutions are known by how much and how well they can account for a broad range of consequences.

References

Bradley, B. A. (2009). Seismic hazard epistemic uncertainty in the San Francisco bay area and its role in performance-based assessment. *Earthquake Spectra, 25*(4), 733–753.

Cho, S., et al. (2006). Calibration of default bridge-damage model. Appendix K in *REDARS 2 methodology and software for seismic risk analysis of highway systems,* by Werner, S. D., et al. Buffalo: MCEER under FHWA Contract Number DTFH61-98-C-0094, United States Department of Transportation.

Collins, J. D., & Carbon, S. L. (2010, 19–21 May). Launch risk acceptability considering uncertainty, In *Proceedings of the 4th International Association for the Advancement of Space Safety (IAASS) conference*, Huntsville.

Dunn, P. M. (2007). Perinatal lessons from the past: Sir Norman Gregg, ChM, MC, of Sidney (1892–1966) and rubella embryopathy. *Archives of Disease in Childhood Fetal and Neonatal Edition*, 092,: F513–514; Louisiana Office of Public Health, Infectious Disease Epidemiology Section, 2012, "Rubella," 7/02/2012, www.infectiousdisease.dhh.louisiana.gov

Forrest, J. M., Turnbull, F. M., Sholler, G. M., Hawker, R. E., Martin, F. J., Burgess, M. A., & Doran, T. T. (2002). Gregg's congenital rubella patients 760 years later. *Medical Journal of Australia, 177*(11), 664–667.

Golub, A. A., & Markandya, A. (2008). *Modeling environment—Improving technological innovations under uncertainty*. London: Routledge, Taylor & Francis Group.

Gower, T. (2002). A history of fluoride, In Prevention. http://www.prevention.com Medical discovery, "Trustworthy endodontist," Accessed 12 June 2013. http://www.discoveriesinmedicine.com/Enz-Ho/Fluoride-Treatment-Dental.html. Accessed 12 June 2013.

Graf, W. P., & Lee, Y. (2009, February). Code-oriented damage assessment for buildings. *Earthquake Spectra, 25*(1), 17–37.

Harre, R. (1983). *Great scientific experiments: Twenty experiments that changed our view of the world*. Oxford: Oxford University Press.

Kagan, Y. Y., et al. (2007, January/February). A testable five-year forecast of moderate and large earthquakes in Southern California based on smoothed seismicity. *Seismological Research Letters, 78*(1), 94–98.

Kraut, A. M., Dr. Joseph Goldberger & the War on Pellagra. http://history.nih.gov/exhibits/goldberger/index.html

Lakatos, I. (1978). *The methodology of scientific research programmes*. London: Cambridge University Press.

Lempert, R. J., Popper, S. W., & Bankes, S. C. (2003). *Shaping the next one hundred years: New methods for quantitative, long-term policy analysis*. Santa Monica: RAND.

Lempert, R. J., Groves, D. G., Popper, S. W., & Bankes, S. C. (2006, April). A general, analytic method for generating robust strategies and narrative scenarios. *Management Science, 52*(4), 514.

Levy, H. (2006). *Stochastic dominance: Investment decision making under uncertainty* (2nd ed.). New York: Springer.

Maloney, W. J., & Maura, M. (2009, Spring). Dr. Frederick McKay: Father of communal fluoridation. *Journal of the Massachusetts Dental Society, 58*(1), 32.

Markowitz, H. M. (1959). *Portfolio selection: Efficient diversification of investments*. Oxford: Basil Blackwell Ltd.

Murnane, R. J., Taylor, C. E., Jagger, T., & Hu, Z. (2011). Robust simulation for sensitivity analysis of catastrophe risk losses. In M. H. Faber, J. Koehler, & K. Nishijima (Eds.), *Applications of statistics and probability in civil engineering* (pp. 875–877). New York: CRC Press.

National Research Council (NRC). (1997). *Review of recommendations for probabilistic seismic hazard analysis: Guidance on uncertainty and use of experts*. Washington, DC: National Academy Press.

NIH, National Institute of Dental and craniofacial Research, *The story of fluoridation*. http://nider.nih.gov/OralHealth/Topics/Fluoride/TheStoryofFluoridation.htm

Perkins, D. (2002). Uncertainty in probabilistic seismic hazard analysis. In C. Taylor & E. VanMarcke (Eds.), *Acceptable risk processes: Lifelines and natural hazards*. Reston: American Society of Civil Engineers.

Petersen, M. D., Cao, T., Campbell, K. W., & Frankel, A. D. (2007, January/February). Time-independent and time-dependent seismic hazard assessment for the state of California: Uniform California earthquake rupture forecast model 1.0. *Seismological Research Letters, 78*(1), 99.

Petersen, M. D., Frankel, A. D., Harmsen, S. C., Mueller, C. S., Huller, K. M., Wheeler, R. L., Wesson, R. L., Zeng, Y., Boyd, O. S., Perkins, D. M., Luco, N., Field, E. H., Wills, C. J., & Rukstales, K. S. (2008). Documentation for the 2008 update of the United States national seismic hazard maps, U. S. Department of the Interior, U. S. Geological Survey, Open-File Report 2008–1128.

Resnick, S. (2007). *Heavy-tail phenomena*. New York: Springer.

Riedel, S. (2005). Edward Jenner and the history of smallpox and vaccination. *BUMC Proceedings, 18*, 21–25.

Robinson, V. (1912). Pathfinds in medicine: Semmelwise, the obstetrician. *Medical Review of Reviews, 18*, 232–245.

Rosakis, A. J., Kanamori, H., & Xia, K. (2006). Laboratory earthquakes. *International Journal of Fracture, 138*, 211–218. Springer.

Rose, A., & Liao, S. (2005). Modeling regional economic resilience to disasters: A computable general equilibrium analysis of water service disruptions. *Journal of Regional Science, 45*(1), 75–112.

Silver, N. (2012). *The signal and the noise: Why so many predictions fail—But some don't*. New York: The Penguin Press.

Stewart, J. P., Archuleta, R. J., & Power, M. S. (2008, February). Special issue on the next generation attenuation project, *Earthquake Spectra, 24*(1), 1–341.

Strasser, F. O., Abrahamson, N. A., & Bommer, J. J. (2009). Sigma: Issues, insights, and challenges. *Seismological Research Letters, 80*(1), 32.

Taylor, C., Rix, G., & Liu, F. (2009, December 1). Exploring financial decision-making approaches for use in earthquake risk decision processes for ports, *Journal of Infrastructure Systems, 15*(4), 406–416.

Taylor, C., Lee, Y., Graf, W., Hu, Z., & Huyck, C. (2010). Robust simulation and cat diagnostics for treating uncertainties in catastrophe risk analysis. In J. Li, Y.-G. Zhao, J. Chen, & Y. Peng (Eds.), *Reliability engineering and risk management: Proceedings of the international symposium on reliability engineering and risk management*. Shanghai: Tongji University Press.

Taylor, C., Graf, W., Lee, Y. J., Huyck, C., & Hu, Z. (2011). Sample treatment of uncertainties in earthquake portfolio risk analysis. In B. M. Ayyub (Ed.), *Vulnerability, uncertainty, and risk: Analysis, modeling, and management, proceedings of the ICVRAM 2011 and ISUMA 2011 conferences*. Reston: American Society of Civil Engineers.

Taylor, C., Graf, W., Yajie (Jerry) Lee, Huyck, C., & Hu, Z. (2012). Propagation of uncertainties through robust simulation and future research, In K. K. Phook, M. Beer, S. T. Quek & S. D. Pang (Eds.), *Fifth Asian-pacific symposium on structural reliability and its applications (5APSSRA)*. Singapore: National University of Singapore.

Taylor, C., Murnane, R., Graf, W., & Lee, Y. (2013, February). Epistemic uncertainty, rival models, and closure. *Natural Hazards Review, 14*(1), 42.

Toulmin, S. (1992). *Cosmopolis: The hidden agenda of modernity*. Chicago: University of Chicago Press.

Trifunac, M. D. (1997). Stresses and intermediate frequencies of strong earthquake accelerations. *Geofizika, 14*, 1–27.

Vachon, D. (2005, May and June). Doctor John Snow blames water pollution for cholera epidemic, *Old News, 16*(8), 8–10; UCLA, "Competing theories of cholera". Accessed 12 Mar 2013.

Wald, D. J., et al. (1996, February). The slip history of the 1994 Northridge, California, earthquake determined from strong-motion, teleseismic, GPS, and leveling data. *BSSA, 86*(1B), S49.

Wesson, R. L., Petersen, M. D., et al. (2004). Losses to single-family housing from ground motions in the 1994 Northridge, California, earthquake. *Earthquake Spectra, 20*(3), 1021–1045.

Wittgenstein, L. (1967). *Remarks on the foundations of mathematics*. Cambridge, MA: The M. I. T. Press, first published in 1956, edited by G. H. von Wright, R. Rhees, G. E. M. Anscombe, and translated by G. E. M. Anscombe.

Chapter 8
Modern Thinking, Ensembles, and the "So What" Question

The mountain trees invite their own cutting down; lamp oil invites its own burning up. Cinnamon bark can be eaten; therefore the tree is cut down. Lacquer can be used; there the tree is scrapped. All men know the utility of useful things; but they do not know the utility of futility. (p. 135 in Laotse, The Wisdom of Laotse, *translated by Lin Yutang, New York: the Modern Library, first edition, 1948)*

All things are exchangeable for fire and fire for all things, like gold for goods and goods for gold, or so sings old. (From "Herakleitos," pp. 6–7 in Eight Objects, *by Robert Bringhurst, San Francisco: the Kanchenjunga Press, 1975)*

Abstract This chapter asks the "so what" question: even if one assumes that robust simulation and nonlinear reasoning assist in helping us to understand a variety of phenomena, how does this help in making decisions on risk evaluations as they impact complex systems? The chapter begins with a discussion of the very large number or even in some cases the plethora of qualitative criteria that may be used in major decisions. These pertain to social, political, efficiency, effectiveness, technical, administrative, aesthetic, equity, legal, regulatory, safety, cultural, organization, medical, and educational criteria, to mention a few. On the quantitative side, prospective criteria include benefit-cost, least total mean cost, mean and variance, and additional dynamic financial analysis procedures such as stochastic dominance, almost stochastic dominance, real options, the principle of least regret, and big bet or variability-reduction methods. Given robust simulation outcomes, these may produce diverse results that decision-makers may use. The consideration at present as to which quantitative methods are superior remains under study.

C.E. Taylor, *Robust Simulation for Mega-Risks*, DOI 10.1007/978-3-319-19413-4_8

8.1 Introduction: Managing Expectations, Qualitative Considerations, and Quantitative Decision Procedures

The previous chapters are intended to show a shift in mega-risk studies from a single-solution approach to one that can yield an ensemble of risk estimates. However, the "so what?" question remains. This chapter outlines only very briefly (a) some of the very qualitative features in decision-making and (b) some key quantitative decision procedures that deserve to be explored when decision-makers are faced with an ensemble of risk estimates.

For a great many decisions pertaining to mega-risks, the answers may be relatively easy. These are decisions that might have immediate or obvious widespread benefits. However, for major decisions, there may be significant downsides as well as upsides, and the decisions may require more deliberation. No single criterion may simply provide the answer. The approach to decision-making in such cases is supported by, for instance, views of Baruch Fischhoff and others in which complicated acceptable risk decisions may have many recommendations but no single criterion that provides the answer in all cases.[1] In his *Reason and the Common Good*, Arthur E. Murphy maintains a more reflective view opposed to the:

> parochial dogmatism which arises from an identification of local orthodoxies with universal truth, and a consequent inability to do justice to these moral insights or ideas that fall outside the limits of accredited preconceptions and linguistic properties.[2]

Abjuring simple solutions to complex decision issues, James D. Wallace provides a nonlinear account of reasoning that involves moral or valuation conflicts: "Efforts to solve moral relevance and conflict inevitably change morality."[3]

In view of these mature views of reasoning about risks, the common good, and moral or decision conflicts, this chapter does not purport to arrive at a single criterion for mega-risk decision-making. Instead, this chapter begins with a very brief account of the systems and many qualitative considerations involved in mega-risk decisions. Afterward, there is an account of some prominent quantitative decision procedures that can and have been used in mega-risk decisions.

---

[1] See Fischhoff, Baruch, Sarah Lichtenstein, Paul Slovic, Stephen L. Derby, and Ralph L. Keeney, *Acceptable* Risk, 1981, Cambridge: Cambridge University Press.

[2] From p. 33 in Murphy, Arthur E., 1963, *Reason and the common good: selected essay of Arthur E. Murphy*, edited by William H. Hay, Marcus G. Singer, and Arthur E. Murphy, Englewood Cliffs, N. J.: Prentice-Hall, Inc.

[3] From Wallace, James D., 1988, *Moral Relevance and Moral Conflict*, Ithaca, N. Y.: Cornell University Press

8.2 Complex Systems Implicated in Mega-Risks and a Sample of Qualitative Decision Considerations

Since mega-risks involve social losses and their ameliorations often require social costs, these risks cover numerous potential systems including those that are political, legal, social, cultural, religious, administrative, organizational, environmental, medical, technological, scientific, educational, and regulatory. Some subject matters appear to have predominantly qualitative approaches: poetry, painting, short stories, novels, rhetoric, and lawyering.

Yet, the major problem addressed in these essays is the emergence of multiple interpretations of the technical or risk findings, findings that can come from scientific, engineering, economic, financial, and other quantitative studies. This emergence is different from the view that in qualitative areas, there should ultimately be a convergence in outcomes, so that decision-makers need only to consider one set of risk profiles. Most studies, especially those with fewer resources for development, are likely to contain such single solutions—along with suitable caveats. However, robust simulation (and during most periods nonlinear research issues) has multiple solutions that decision-makers must address. With robust simulation, one cannot properly say, for instance, "Here is the 90th centile confidence interval for losses." Confidence intervals vanish "in the cloud" as the number of simulations increase—as long as the distributions in question have finite variances (alternatively, those supporting alpha distributions and other extreme value distributions will be imposing on the data their supposed centile estimates with very different meanings from those traditionally deployed.)

8.3 Quantitative Decision Procedures of Interest

First deterministic quantitative decision procedures are discussed. Of special interest is the minimax or principle of least regret. Second, a variety of stochastic decision procedures are discussed. These include those stressing arithmetic means (e.g., benefit-cost, least total mean cost), those stressing as well statistical variances (e.g., mean-variance procedures), those stressing entire loss (or loss and gain) distributions (e.g., stochastic dominance, almost stochastic dominance), and those stressing hedging techniques (e.g., real options). Quantitative presentations focus on major outlines of these methods. Full-scale applications to robust simulation techniques require research beyond these essays.

8.3.1 Deterministic Quantitative Decision Procedures of Interest

Beginning at least early in the twentieth century, deterministic decision procedures became of great interest. These assume that one is deciding among known alternatives. Deterministic decision procedures include the principle of dominance, the maximin principle, the minimax principle or principle of least regret, and the combined minimax and maximin principle (or the optimism-pessimism principle).[4]

One major issue with respect to deterministic decision procedures is how to handle initial outlays. Relative to prospective disasters, deterministic procedures are basically procedures used without respect to probabilities Mega-risk scenarios are not certain. Decision alternatives may have fairly certain costs or initial outlays, but not certain benefits. So, one way to add initial outlays to downstream losses is first of all to treat both in terms of constant dollar values. This, however, does not solve the entire major issue inasmuch as the mega-risk scenarios postulated may not occur. Moreover, even if one uses constant dollar values to express losses, losses that occur downstream are subjected in economic terms to discounting. Discount rates are, say, 4 % for federal projects and may be slightly higher for local municipalities and private utilities. (Note that until recently, these discount rates were 7 %. However, current discount rates used are roughly the average difference between government bonds and inflation.[5]) Nonetheless, downstream losses are not—without discounting—directly commensurable with initial outlays.

Deterministic procedures are discussed first with emphasis on the mini-max or "principle of least regret". According to Luce and Raiffa (1957, p. 2), John von Neumann apparently held that the minimax theorem is necessary for the theory of games (a slightly different topic but a large subject matter).[6] According to NASA, because of its "worst-case" feature, this rule had found some application in military systems.[7] According to Kleindorfer et al. this is a widely used principle.[8] In management theory, James March and others have found that many administrators use such a principle.[9] Of special interest to these essays is that the minimax principle has been used in some climate change literature including some of the pioneering work on robust simulation in a 100-year economic forecasting.[10]

---

[4]A very satisfactory account of these and other deterministic principles is found in Resnick, Michael D., 1987, *Choices: An Introduction to Decision Theory*, Minneapolis: University of Minnesota Press.

[5]Rose, Adam, et al., 2007, "Benefit-Cost Analysis of FEMA Hazard Mitigation Grants," *Natural Hazards Review*, November. The use of a discount rate of slightly above 2 % is found in Lempert, R. J., Popper, S. W., and Bankes, S. C., 2003, *Shaping the Next One Hundred Years: New Methods for Quantitative, Long-Term Policy Analysis*, Santa Monica, CA: RAND.

[6]See p. 2 in Luce, R. Duncan and Howard Raiffa, 1957, *Games and Decisions: Introduction and Critical Survey*, New York: Dover Publications, Inc.

[7]See p. 77, National Aeronautical and Space Administration (NASA), 1995, *NASA Systems Engineering Handbook*, Washington D. C.: NASA.

[8]See p. 153ff. in Kleindorfer, Paul R., Howard C. Kunreuther, and Paul J. H. Shoemaker, 1993, *Decision Sciences: An Integrative Perspective*, New York: Cambridge University Press.

[9]In March, James G., 1988, *Decisions and Organizations*, Oxford: Basil Blackwell Ltd., James March and others have found that many administrators use such of principle.

[10]See the American Society of Civil Engineers (ASCE), Committee on Adaptation to a Changing Climate (CACC), 2013, *Bridging the Gap between Climate change Science and Civil Engineering Practice*, edited by J. Rolf Olsen, with many contributing authors, review draft, 2013. This draft builds on, for instance, Lempert, R. J. et al., 2003, op.cit., and Lempert, R. J, Groves, D. G.,

One set of quantitative procedures to characterize the minimax principle is as follows with reference principally to monetary measures:

Let $A_1, A_2, \ldots, A_m$ be m seismic decision alternatives with S_1 representing the baseline or status quo. Associated with each seismic decision alternative is a marginal cost $C_1, C_2, \ldots, C_n$, respectively. The marginal cost represents the present value of the proposed upgrade or upgrades for the decision alternative. For many purposes, this can be called the initial outlay. **For other purposes, maintenance costs will be critical.** Note that $C_1 = 0$ for the baseline or status quo.

Let $S_{i1}, S_{i2}, \ldots S_{ij}, \ldots, S_{in}$ be n system states (simulations) defined for alternative A_i. In a matrix of alternatives and system states, S_{ij} will be the system loss for the *i*th row (or alternative) and the *j*th column (or scenario simulation, typically called "state" or "system state").

For these reasons, it is essential to add a scenario—no costs. On this scenario, the total costs are merely the initial outlays, namely, C1, C2, ..., Cn, respectively. Within the time horizon for decision-making, this may or may not be a plausible assumption.

For this purpose, we shall let

$$S_y' = C + Sy \tag{8.1}$$

where S_y' is the system state for *Alternative A* that considers the initial outlays.

In effect, for each of the decision alternatives, one needs to pick a plausible least dollar loss scenario within the time horizon in question and add this least cost scenario to the cost for the decision alternative. One could use stochastic methods in order to derive a least loss scenario for the time horizon used. Alternatively, one may assume more simply that zero is the least loss scenario for the time horizon.

This requirement that there be a zero loss scenario within the time frame of decision-making follows that dictum for potential disasters: "certain costs but uncertain benefits." This dictum emphasizes that costs of the seismic decision alternatives are by and large certain (to the extent that construction and other costs are certain), whereas benefits may range from very large to none at all.

For example, according to Table 8.1, there are four scenarios (simulations, system states) selected, and total system losses (including initial outlays) for alternative A_1 have a maximum of \$9M, whereas total system losses (including initial outlays) for alternative A_2 have a maximum of \$10M.

Table 8.1 Illustrative table of two alternatives and four total system losses

	Scenario losses (plus initial outlays)			
Alternative A_1	\$3M	\$7M	\$9M	\$9M
Alternative A_2	\$0M	\$8M	\$2M	\$10M

Popper, S. W., and Bankes, S. C., 2006, "A General, Analytic Method for Generating Robust Strategies and Narrative Scenarios," *Management Science*, Vol. 52, No. 4, April, pp. 514–528.

The minimax principle basically says that one should undertake an action that will involve the least regret (greatest relative gain) for some one scenario.

The formulas for the minimax regret rule are somewhat more complicated than those for other deterministic principles. In particular—and put in terms of losses rather than gains—recall that S_{ij} represents the total system loss for the ith row, alternative I, and the jth simulation for the seismic decision alternative i:

Define

$$M_j = \text{minimum}\{S'_{ij}\} \text{ for all } i \tag{8.2}$$

and

$$Rij = S_{ij} - M_j \text{ for each } i, j \tag{8.3}$$

$$\text{maxregret}_i = \max R_{ij} \text{ for each } i, j \tag{8.4}$$

The minimax regret rule thus states that

$$\text{Alternative } Aq \text{ is superior to alternative } Ar \text{ if maxregret}_r > \text{maxregret}_q \tag{8.5}$$

This rule can be illustrated again in terms of Table 8.1. In Table 8.1, and using Eq. (8.2), one derives:

$$M_1 = \$0M$$
$$M_2 = \$7M$$
$$M_3 = \$2M$$
$$M_j = \$9M$$

From these values, Table 8.1, and Eq. (8.3), one derives the following regret Table 8.2.

Using Eq. (8.4), one then shows that the maximum regret for alternative A_1 is \$7M, whereas the maximum regret for alternative A_2 is \$1M. Thus, according to the rule expressed in Eq. (8.5), the minimax regret rule favors alternative A_2.

One of Michael Resnick's criticisms of the minimax regret rule is that the addition of alternatives may reverse the outcome of the decision rule. To this end, Table 8.3 has been constructed. If one uses Eqs. (8.2), (8.3), (8.4), and (8.5), then one derives that Alternative A_1 is superior to Alternative A_2, which is in turn superior to Alternative A_3. The maximum regrets are \$7M, \$8M, and \$9M, respectively, for the alternatives in order. Thus, the addition of a new alternative (in this case Alternative

Table 8.2 Regret table derived from Table 8.1

	Regrets (R_{ij}'s in Eq. (8.3))			
Alternative A_1	\$3M	\$0M	\$7M	\$0M
Alternative A_2	\$0M	\$1M	\$0M	\$1M

Table 8.3 Illustrative table of three alternatives and four system losses

	Scenario losses (plus initial outlays)			
Alternative A_1	\$3M	\$7M	\$9M	\$9M
Alternative A_2	\$0M	\$8M	\$2M	\$10M
Alternative A_3	\$4M	\$0M	\$6M	\$18M

Table 8.4 A second illustrative table of two alternatives and four system losses

	Scenario losses (plus initial outlays)			
Alternative A_1	\$8M	\$7M	\$9M	\$0M
Alternative A_2	\$1M	\$0M	\$2M	\$8M

Table 8.5 Example of a scenario table for insurance purchase

	No earthquake damage	Earthquake damage
No insurance	\$0.0	\$1M
Insurance	\$0.2M	\$0.2M

A_3 to the other two alternatives in Table 8.1) can indeed reverse the outcome based on the minimax regret rule.

Resnick provides an additional criticism that there are cases in which the minimax regret rule yields very counterintuitive conclusions.[11] To illustrate, in Table 8.4, the minimax regret rule would favor Alternative A_1 over Alternative A_2 even though Alternative A_1 is clearly the worse alternative in three of the four scenarios. Table 8.4 could be extended to a great many cases in which Alternative A_1 is \$7M or so worse than Alternative A_2, but the \$8M difference in one scenario is driving the decision, regardless of these other cases.

The principle of least regret can sometimes lead to the equivalent of "high stakes gambling." Table 8.4 (especially when extended) can illustrate how this may occur. In effect, alternative A_1 can be constructed to be a steady loser (one can fill in additional scenarios in which losses are similar to those in the first three columns). However, there can be one scenario in which the relative gains of alternative A_1 over alternative A_2 can exceed the relative gains of alternative A_2 over alternative A_1 for the rest of the scenarios. Thus, selection of alternative A_1 over Alternative A_2 can reflect the tendency to go for the largest prize (relative gain) in spite of all the other losses that are expected to accrue as a result of the decision.

Similar criticisms obtain when one considers insurance purchase. Modifying an illustration from Kleindorfer,[12] one may develop a regret table as follows for insurance purchase:

From Table 8.5, one can derive a regret table that shows that one would most regret the alternative "No insurance." Yet, such a regret table fails to do justice to the

[11] In M. Resnick, 1987, op. cit., the minimax rule is discussed on pp. 28–32.

[12] See p. 153ff. in Kleindorfer, Paul R., Howard C. Kunreuther, and Paul J. H. Shoemaker, 1993, *Decision Sciences: An Integrative Perspective*, New York: Cambridge University Press.

cost of insurance relative to its probable benefits, the need to renew insurance periodically (annually) with its total downstream cost not being its singular cost in one context, and a myriad of other considerations in this context (e.g., the presence or absence of state and federal post-disaster assistance programs).

In spite of these limitations, the minimax principle has been used in association with robust simulation results to provide useful results, as in the case of nascent work by R. Lempert and others. The presence of limitations on quantitative principles is consistent with the remarks on reasoning about the common good and moral and other conflicts at the outset of this chapter.

8.4 Probabilistic Quantitative Decision Procedures of Interest

8.4.1 Probabilistic Quantitative Principles Stressing Averages Alone

Among those quantitative decision principles that focus on averages or central values are benefit-cost methods, least total mean costs, and the use of logic tree weights on robust simulation results in order to assess central values for decisions.

8.4.1.1 Benefit-Cost Methods

Benefit-cost methods are most suitably used for governments and possibly huge organizations that are self-insured or for whom the probability of default is virtually nil. These organizations thus do not need to worry about the "volatility" or variability in costs and benefits. State and local governments do not comprise such organizations nor do the vast number of private sector organizations.

First, benefits are the present value of reduced losses. Let us suppose that there are Y years of random walks for these n scenarios. Then,

$$\text{Expected annualized status quo losses} = \frac{S_{11} + S_{12} + \cdots + S_{1j} \cdots + S_{1n}}{Y} \tag{8.6}$$

Moreover,

$$\text{Expected annualized losses for Alternative } i = \frac{S_{i1} + S_{i2} + \cdots + S_{ij} \cdots + S_{in}}{Y} \tag{8.7}$$

$$\begin{aligned}&\text{Expected annualized benefits for alternative}\\ &i = \text{Expected annualized status quo losses -}\\ &\text{Expected annualized losses for alternative } i\end{aligned} \tag{8.8}$$

If D is the discount rate selected, and t is the number of years for the assumed life-span of the system modifications, then

$$\text{The present value multiplier } PVM = \frac{1-(1+D)^{-t}}{D} \tag{8.9}$$

From these equations, it follows that

The benefits of alternative
$i = PVM *$ (Expected annualized benefits for alternative i) (8.10)

From these equations, it follows that

For benefit-cost evaluations,
Alternative i is preferred to the status quo if
the benefits of alternative I exceed its costs (8.11)

One can also compare alternative i to other alternatives, which leads into the least cost evaluation.

Note that if for each robust simulation outcome the resulting distribution yields a favorable benefit-cost ratio for some alternative, and this alternative is favored over all alternatives, then this definitely favors the superior alternative. However, variability in outcomes is also desirable for the vast number of organizations and governmental entities.

8.4.1.2 Least Total Average Costs Methods

With respect to mean values of losses, benefit-cost evaluations do not necessarily lead to an optimum solution. Instead, given a benefit-cost ratio exceeding unity, the evaluation is presumed to be complete. Such is not the case with least cost methods, which extend benefit-cost methods to all decision alternatives evaluated. In particular, one defines

The otal (mean) costs of alternative $i = TMCi =$
$Ci + PVM *$ (Expected annualized losses for alternative i) (8.12)

And then the rule for least total (mean) costs becomes

Alternative i is preferred to alternative j if $TMCi < TMCj$ (8.13)

One can see that this least total costs method is a way of comparing all alternatives against each other, and so arriving at the alternative that has the best benefit-cost ratio. In that sense, the least total costs method optimizes with respect to mean losses and costs, whereas benefit-cost methods do not necessarily optimize with respect to mean losses and costs.

Unlike the benefit-cost approach, the least total mean approach is likely to yield as many "optimal" outcomes as there are robust simulation approaches. Decision-makers would then need to have additional criteria to select among outcomes.

8.4.1.3 Subjective Weighting of Average Costs and Benefits

Here decision-makers may have preferences for one or more of the robust simulation approaches and may decide to weight each of the approaches. Using this method, one can derive from the average outcomes for each robust simulation approach a weighted average of benefit-cost ratios. Likewise, one could use this weighted simulation approach method in order to select the average weighted "optimal" approach.

The "weighted" approach discussed here should not be described as a method for determining robust simulation approaches and outcomes themselves, although it may be challenging to distinguish the "subjective" elements in critical studies for the more "objective" methods.

8.5 Stochastic Approaches Considering Variability as well as Averages

8.5.1 Mean-Variance Decision Methods

The mean-variance method has been widely used for over 20 years in the capital markets in order to assure that investors are sufficiently diversified. In particular, the mean-variance criterion evaluates investments not only with respect to their mean yields but also with respect to their variance, measured by their statistical variance (or, for practical purposes equivalently, their standard deviation). This method is especially suitable for projects by entities that are not extraordinarily diversified, such as individuals, corporations, and state and local governments.[13]

For the application of the mean-variance rule, one first calculates the variance of total mean costs, or

$$\mathrm{Var}_i(TMC) = \left(\sum_{j=1}^{n} (TMCi - Sij) \right) \Big/ (n-1) \tag{8.14}$$

[13] See Markowitz, H. M., 1959, *Portfolio Selection: Efficient Diversification of Investments*, Oxford: Basil Blackwell Ltd.; Levy, Haim and Marshall Sarnat, 1984, *Portfolio and Investment Selection: Theory and Practice*, New Jersey: Prentice Hall International; and p. 236 in Bernstein, Peter, 1996, *Against the Gods: The Remarkable Story of Risk*, New York: John Wiley & Sons, Inc.

For decision purposes one may equivalently calculate the standard deviation of the total mean costs since the standard deviation is on a quantitative scale more commensurate with the total mean cost:

$$STDi(TMC) = [\mathrm{Var}_i(TMC)]^{0.5} \tag{8.15}$$

The mean-variance criterion is thus expressed as follows:

$$\begin{aligned}&\text{Alternative } i \text{ is preferable to alternative } j \text{ if}\\ &\text{both } TMC_i < TMC_j \text{ and } \mathrm{Var}_{i(TMC)} < \mathrm{Var}_{j(TMC)};\\ &\text{If neither alternative } i \text{ is preferable to alternative } j \text{ nor}\\ &\text{alternative } j \text{ is preferable to alternative } i,\\ &\text{then they are indifferent to each other}\end{aligned} \tag{8.16}$$

In the above formulations, the standard deviation of total mean costs may replace the variance with of course no difference in the decision outcomes. Owing to scaling issues, this reformulation of the rule in (8.16) is probably desirable if for a set of alternatives total mean costs and their standard deviations are either table or plotted.

The mean-variance rule is thus more restrictive than the least cost or benefit-cost rules. In other words, the mean-variance rule takes into account two dimensions of the investment decision: not only the mean total costs but the "volatility" (as measured by the statistical variance) of the decision. Thus, fewer decision alternatives may be ruled out by the mean-variance criterion than by the least total (mean) cost criterion.

Note that the principle of dominance can help to indicate some instances (counterexamples) in which the mean-variance criteria do not apply well. In the first instance, there are problems with riskless assets, those in which one puts monies under the bed or in which one invests in virtually riskless federal securities. Take, for example, the following two equally probable returns on alternatives:

Alternative 1	1	2	1	2
Alternative 2	1	1	1	1

Now, the principle of dominance would imply that alternative 1 is superior to the riskless alternative 2. However, since the riskless alternative has a variance of zero, even though its mean return is less than that of alternative 1, the MVC implies that neither is preferred to the other.

A similar example arises from Fishburn and Vickson[14] for the following two alternatives:

[14] See p. 62, Fishburn, P. C. and Vickson, R. G., 1978. "Theoretical Foundations of Stochastic Dominance." *Stochastic Dominance. An Approach to Decision-Making Under Risk*, G. A. Whitmore, and M. C. Findley, eds., D. C. Heath, Lexington, MA, 39–114

Alternative 1	1	1	4	4	4	4
Alternative 2	1	1	3	3	4	4

The principle of dominance implies that alternative 1 is superior to alternative 2. However, using the MVC, one would need to calculate the variances in order to determine whether indeed alternative 1 is preferable.

Even without these counterexamples, the MVC approach has come under great criticisms from the failure of Long-Term Capital Management, faulty applications that assisted in causing the Great Recession of 2008 and in general limitations on the approach when it is used beyond Gaussian or very linear or quadratic assumptions.[15] The use of averages for financial and general economic evaluations would have had even more deleterious effects.

8.5.2 Stochastic Dominance and Almost Stochastic Dominance Approaches

A full discussion of stochastic dominance and almost stochastic dominance approaches are found in Levy.[16] A discussion of how these apply to seismic design and redesign (and similar) issues is found in Taylor, Rix, and Liu.[17] In effect, first-order stochastic dominance applies in those cases in which one alternative is superior to another alternative in at least one instance (simulation) and the first alternative is at least equal to the second alternative in all other cases. The study of seismic design issues has led to the conclusion that second-order stochastic dominance does not show that superior design is better in cases in which there are some simulations (e.g., 50-year random walks) in which no damage would have occurred to either the superior or inferior design. For instance, no damaging earthquake has occurred in the city of Orange, California, since the 1933 Long Beach earthquake. Thus, a building design in a high seismicity region need not exhibit any benefits in all 50-year life spans—and even though seismic design costs tend to be well below seismic retrofit costs.

Still, second-order stochastic dominance and almost stochastic dominance approaches appear to be viable tools for decision-making in the context of robust simulations for mega-risks.

---

[15] Samples of criticisms of the MVC approach are found in Lowenstein, R., 2000, *When Genius Failed: The Rise and Fall of Long-Term Capital Management*, New York: Random House; Levy, H., 2006, *Stochastic Dominance: Investment Decision Making Under Uncertainty*, 2nd edition, New York, NY: Springer; and Taleb, Nassim Nicholas, 2012, *Antifragile: Things that Gain from Disorder*, New York: Random House.

[16] Levy, H. 2006, ibid.

[17] See Taylor, Craig, Glenn Rix and Fang Liu, 2009, "Exploring Financial Decision-Making Approaches for Use in Earthquake Risk Decision Processes for Ports," *Journal of Infrastructure Systems*, volume 15, number 4, pp. 406–416, December 1, 2009.

8.5.3 Hedging: Real Options

Real options are those in which not all monies are spent initially in order to treat a mega-risk. These are discussed, for instance, in Lempert et al. and in Golub and Markandrya.[18] These tools are especially valuable when (a) waiting until a later date does not engender a catastrophe or enable a condition to be serious enough to be a likely cause of catastrophe and (b) more can be learned from more limited treatments used and as the science and technology improves.

8.5.4 Hedging with Anti-fragile and/or Big Bets Reduction Methods

So far in this chapter, there has been no reference to findings in Chap. 5 in which distributions evaluated may turn out to be nearly are actually extreme, and/or there may be a high probability of ruin or catastrophe. Work on hedging to avoid or reduce cost-effectively variability and/or big (and somewhat unstable) bets are two ways in which one may approach mega-risks that appear without treatments to be nearly or actually extreme.

References

American Society of Civil Engineers (ASCE). Committee on Adaptation to a Changing Climate (CACC). (2013). *Bridging the gap between climate change science and civil engineering practice* (J. Rolf Olsen, Ed.), Reston, VA.

Bernstein, P. (1996). *Against the Gods: The remarkable story of risk*. New York: Wiley.

Fischhoff, B., Lichtenstein, S., Slovic, P., Derby, S. L., & Keeney, R. L. (1981). *Acceptable risk*. Cambridge: Cambridge University Press.

Fishburn, P. C., & Vickson, R. G. (1978). Theoretical foundations of stochastic dominance. In G. A. Whitmore & M. C. Findley (Eds.), *Stochastic dominance. An approach to decision-making under risk* (pp. 39–114). Lexington: D. C. Heath.

Golub, A. A., & Markandya, A. (2008). *Modeling environment—Improving technological innovations under uncertainty*. London: Routledge, Taylor & Francis Group.

Herakleitos. (1975). *Eight objects, by Robert Bringhurst*. San Francisco: The Kanchenjunga Press.

Kleindorfer, P. R., Kunreuther, H. C., & Shoemaker, P. J. H. (1993). *Decision sciences: An integrative perspective*. New York: Cambridge University Press.

Laotse. (1948). *The wisdom of Laotse* (L. Yutang, Trans., 1st ed.). New York: The Modern Library.

Lempert, R. J., et al. (2003a, April). *op.cit.,* and Lempert, R. J, Groves, D. G., Popper, S. W., Bankes, S. C. (2006). A general, analytic method for generating robust strategies and narrative scenarios. *Management Science*, *52*(4): 514–528.

[18] See Lempert, R. J., 2003, op. cit., and Golub, Alexander A. and Anil Markandya, 2008, *Modeling Environment—Improving Technological Innovations Under Uncertainty*, London: Routledge, Taylor & Francis Group.

Lempert, R. J., Popper, S. W., & Bankes, S. C. (2003b). *Shaping the next one hundred years: New methods for quantitative, long-term policy analysis*. Santa Monica: RAND.
Levy, H. (2006). *Stochastic dominance: Investment decision making under uncertainty* (2nd ed.). New York: Springer.
Levy, H., & Sarnat, M. (1984). *Portfolio and investment selection: Theory and practice*. Englewood Cliffs: Prentice Hall International.
Lowenstein, R. (2000). *When genius failed: The rise and fall of long-term capital management*. New York: Random House.
Luce, R. D., & Raiffa, H. (1957). *Games and decisions: Introduction and critical survey*. New York: Dover Publications.
March, J. G. (1988). *Decisions and organizations*. Oxford: Basil Blackwell Ltd.
Markowitz, H. M. (1959). *Portfolio selection: Efficient diversification of investments*. Oxford: Basil Blackwell Ltd.
Murphy, A. E. (1963). In W. H. Hay, M. G. Singer, & A. E. Murphy (Eds.), *Reason and the common good: Selected essay of Arthur E. Murphy*. Englewood Cliffs: Prentice-Hall.
National Aeronautical and Space Administration (NASA). (1995). *NASA systems engineering handbook*. Washington, DC: NASA.
Resnick, M. D. (1987). *Choices: An introduction to decision theory*. Minneapolis: University of Minnesota Press.
Rose, A., et al. (2007, November). Benefit-cost analysis of FEMA hazard mitigation grants. *Natural Hazards Review, 8*(4): 97–111.
Taleb, N. N. (2012). *Antifragile: Things that gain from disorder*. New York: Random House.
Taylor, C., Rix, G., & Liu, F. (2009). Exploring financial decision-making approaches for use in earthquake risk decision processes for ports. *Journal of Infrastructure Systems, 15*(4), 406–416.
Wallace, J. D. (1988). *Moral relevance and moral conflict*. Ithaca: Cornell University Press.

Part V
Possible Futures

Chapter 9
Remaining Questions: Conclusions and Queries

In theology or philosophy, you may (with due intellectual modesty) adopt as personal working positions the ideas of your inherited culture; but you cannot deny others the right to adopt different working positions for themselves, let alone pretend that your experience "proves" the truth of one such set of opinions, and the necessary falsity of all the others. (From p. 29, Toulmin, Stephen, 1992, Cosmopolis: The Hidden Agenda of Modernity, *Chicago: University of Chicago Press)*

The study of the confirmation procedures as they are practiced in the sciences is ... often the study of what scientists will and will not give up in order to gain other particular advantages. (From p. 212 in Kuhn, Thomas, 1977, The Essential Tension: Selected Studies in Scientific Tradition and Change, *Chicago: University of Chicago Press)*

Abstract This chapter attempts to characterize the current "equilibrium" state of thinking with respect to probability and statistics for systems-based problems. This chapter looks at questions still to be addressed: What are the roles of experts and the enormous amount of drill and other convergent approaches in education and training, including the presupposition of "universals" assumed pedagogically? How does this education and training provide a world view that assists in "bridging the gap" between finite samples and infinite populations—even though the bridges are corrigible? How does one deal with such "deterministic" subjects as chaos theory when unknown initial conditions provide room for developing probabilistic models? Does Taleb's contention of the incomputability of Black Swans result from the "wobble" inherent in extreme value distributions, or can they be and have they been successfully applied with qualifications? How does one use nonlinear reasoning to understand differently the so-called "fallacy of affirming the consequent" when so many theories are praised for their successes? This chapter also asks how competition can be encouraged and what decision procedures work best with robust simulation outcomes. This chapter specifically addresses how many of the concepts covered in this book including robust simulation, instabilities in extreme value distributions, and linear reasoning upset a very long-standing Western tradition of believing that there is but a unique solution, a singular truth to be achieved.

C.E. Taylor, *Robust Simulation for Mega-Risks*, DOI 10.1007/978-3-319-19413-4_9

9.1 Further Elaboration of Robust Simulation

Robust simulation has been used to account for uncertainties in the computer generation of loss estimates to portfolios and systems. The use of alternative models provides the basis for calculating the bounds of uncertainties. These bounds do not necessarily represent confidence intervals because not all alternative models are typically considered and the most plausible values may lie at or near the extremes of these bounds. These bounds do not comprise a continuum or a "fuzzy set." As in chaos theory, different initial conditions posited for a modeled system may yield wildly different trajectories or loss estimates. So, for specific loss estimates (e.g., as those with a given return interval), different comprehensive models may yield different estimates. The ensemble of these estimates does not necessarily comprise a continuum.

The use of robust simulation arises because resources are adequate to provide for competitive outlooks, competitive outlooks are available, and single methods and approaches have severe limitations in accounting for uncertainties in loss distributions. The presence of these social and economic resources implies that the value of competition is recognized. A process is in place that permits not a monopoly of results but instead possible different directions in which subject matters may go. This process by diverse investigators does not need to converge as long as there is recognition that one can deal with changes in and diverse interpretations of risk estimates.

The application of deductivist, frequentist, or Bayesian methods to develop confidence intervals does not help insofar as these confidence intervals virtually vanish as the number of trials increases. In addition, the sharp epistemic/aleatory dichotomy does not fit any of these aforementioned approaches, does not fit the current approach being proposed, and results in either arbitrary or self-contradictory results.[1]

The use of confidence intervals does not address this uncertainty issue also because in some cases loss estimates are unstable. As a consequence, catastrophe indexes have been developed in Chap. 5 to provide an "empirical" (finite trial) estimate of the stability of the loss estimates in question. Unstable estimates may be conceived of as "wobbling" indefinitely as individual trials or more often sequences of trials always have the potential to upset an apparently stable estimate.

Robust simulation requires the application of alternative models that are as coherently constructed as possible. The existence of alternative credible models results from the flexibility inherent in probability and statistics to pursue divergent approaches. The nonlinearity of major discoveries as indicated in Chap. 7 often provides considerable temporal gaps between the development of alternative theories and a final resolution—to the extent that one exists.

[1] See Taylor, C., R. Murnane, W. Graf, and Y. Lee, 2013, "Epistemic Uncertainty, Rival Models, and Closure," *Natural Hazards Review*, February, pp. 42–51, volume 14, number 1.

These alternative models are also possible because Bayesian, frequency, and other approaches have considerable but not unconditional cognitive value. Thus, previous chapters do not rule out categorically applications of Bayesian, frequency, or mathematization approaches but instead regard them as having value in a conditional sense.

This conditional nature of statistics can also be shown if one attempts to take an extremely huge database, collected over centuries and diverse circumstances, without properly distinguishing among these circumstances. For instance, one may take life-expectancy data from London in the 1600s and compare this with 1993 US life-expectancy data. Their combination would result in multimodality as in the chart that follows (Fig. 9.1).[2]

Not only do the Bayesian, frequency, and predominantly mathematical approaches rely at best on finite samples (like the finite samples in the partition used for the theorem of total probability) but so do the robust simulation and catastrophe index approaches. The uncertainty bounds generated by a robust simulation approach are restricted to the state-of-the-practice and so may be improved as the alternative credible models are improved.

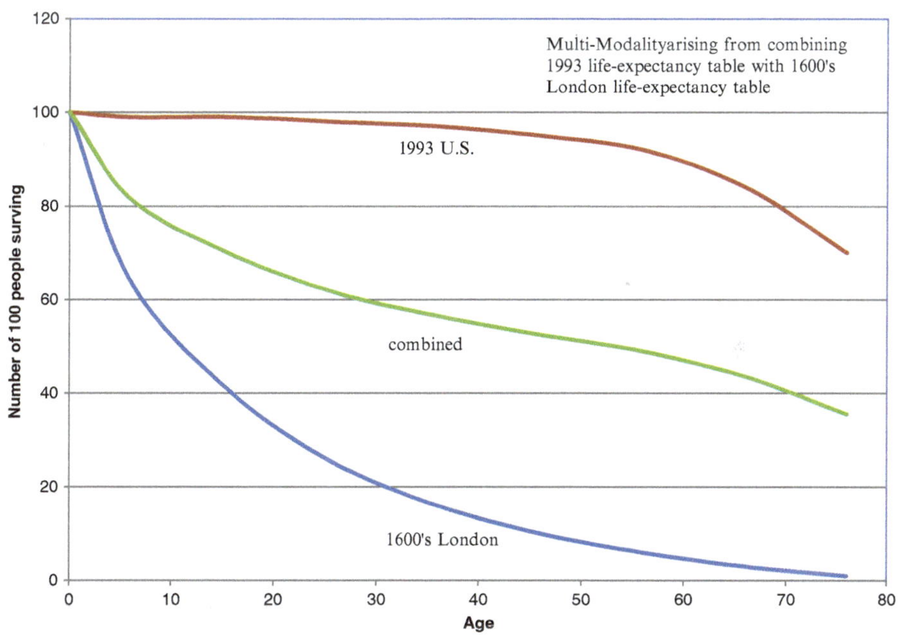

Fig. 9.1 Multimodality arising from combining 1993 life-expectancy table with 1600s London life-expectancy table (x-axis is age; y-axis is number of original 100 in population who have survived to that age)

[2] From p. 83 in Bernstein, Peter L., 1996, *Against the Gods: The Remarkable Story of Risk*, New York: John Wiley & Sons, Inc.

9.2 Tradition of Believing in Only Unique Risk Solutions

According to Stephen Toulmin, the view that inquiry yields a unique result in response to critical questions has prevailed in Western countries for well over three centuries since the seventeenth century and still seems to permeate many quantitative studies today. The development of ecological studies and the interest in some deeply practical studies such as health have provided some limitations on extreme versions of the distinction between theory and practice, still common in Western universities today. The alignment of mathematics with theory has too been part of this Western tradition, and in experimentation and the use of field data, the belief that "almost surely" enables one to assume the convergence of results at infinity is part of this common traditional system of beliefs. Toulmin writes that the "mathematical and experimental" have not been clearly dovetailed in this long-standing tradition.[3] Given that indefinitely many experiments are infinitely costly and otherwise impossible, a different approach to the nexus between quantitative models and experiments needs to be developed. Likewise, "validation" of quantitative models turns out to rest on assumptions that surely extend beyond the bounds of experience.

Absent the view that there is but one quantitative solution to risk evaluations of systems and other similar critical problems, the specter of freedom in inquiry rears its ugly head. Freedom in general is often lauded, but discretion in pursuing alternative approaches to problem-solving is often tolerated only if it is presumed that there is but one truth to the matter in question.

9.2.1 Flexibility in Approaches

The previous discussion by Menke of how diverse measures (L1, L2, and so on) can be used to evaluate "fitting" of a model to data grossly underestimates the mathematical combinations available for "fitting." The measures could include L0.5, L1.3, L2.8, and so on as well as a direct weighting of the tail (as in the catastrophe indexes described in Chap. 5, which ignores the body of the distribution).[4]

The discussion by N. Silver on how an underlying distribution is required for classical fitting techniques as first developed by the biologist R. A. Fisher and others again assumes an infinite number of cases can be "validated" by a finite number of samples. The statistician David Freedman has shown that the use of "percent" acceptability misleads one to believe that this constitutes the probability of accept-

[3] Pertinent pages include pp. 84, 104 Toulmin, Stephen, 1992, op. cit.

[4] On how rational numbers as those referenced must be used for "univariate" alpha distributions in which the slope is less than 2 and in some cases less than 1, see p. 15 in Nolan, John P., 2009, *Stable Distributions: Models for Heavy Tailed Data*, accessed on the Internet 2/27/13.

ability of the "null" hypothesis. Among other things, this acceptability resides only to the extent that the underlying distribution obtains—a huge assumption.

For some time in environmental studies, there have been multiple interpretations of nature and of course of the behavior of people and other living beings. In climate change studies, "there is not one model…there are multiple models, multiple scenarios, the interaction of socioeconomic data, the introduction of parameterizations, nested ecological models, and…the communities interact, ask more questions,… new questions result in different experimentation…new results, new answers, more new questions…."[5]

Chapter 7 has shown that research into major health issues is not necessarily resolved in a very short span of time. Instead, clarification of initial discoveries, or educated guesses based on early findings, can come considerably later. Through these later developments, alternative determinants of specific diseases, for instance, can be discovered. Filtration plants are extremely valuable but may not provide "safe" water if the main piping systems are old. Water may need to be boiled at the destination point in order to be safe. Beriberi may have other forms than the "dry" beriberi for which adequate intake of thiamine tended to be a major cure. Fluoridated water may prove to be valid in reducing tooth decay in a number of cases, but not in those in which alternative sources of fluoride in drinks, food, or toothpaste provide adequate means to reduce decay.

9.3 Five Queries and Partial Current Responses

The results presented raise serious questions for making risk estimates for socially important issues. What follows are five of these questions.

Question 1: **what are the roles of experts and the enormous amount of drill and other convergent approaches in education and training, including the presupposition of "universals" assumed pedagogically**? Why should "professionals" or "experts" be trusted any more than anyone else? Haven't professionals made indefinitely many mistakes, a lot of which have been critical? How does one choose among approaches that are credible and those that are not?

Over time, the notion of "professionals" has of course changed immensely with the changes in universities and diverse disciplines as well as their association with society. Not too long ago, some of the scientists most highly regarded, Charles Darwin for one, were "amateurs." So, the notion of "professional" here has to do with the types of disciplinary studies and investigations undertaken by Darwin or many others who have contributed to critical areas.

The famous historian of science Thomas Kuhn considers the type of "divergence" discussed in these essays to require a substantial "convergence." Dealing princi-

[5] Written communication, Melissa Dresler, 7/22/13.

pally with the nonhistorical sciences (e.g., not, for instance, geology or evolutionary biology), Kuhn maintains that the bulk of a scientist's early career resides in "puzzle solving," normal science, or very convergent thinking. As Kuhn remarks, "it is often better to do one's best with the tools at hand than to pause for contemplation of divergent approaches." In contrast to those who treat creativity, or divergent thinking, as involving isolated discoveries, Kuhn avows that "the ultimate effect of this tradition bound work [puzzle-solving] has been to change the tradition."[6]

Very importantly, for Kuhn science is systemic and hence progressive in the noncumulative sense:

> most new discoveries and theories in the sciences are not merely additions to the existing stockpile of scientific knowledge. To assimilate them the scientist must usually rearrange the intellectual and manipulative equipment he has previously relied upon, discarding some elements of his prior belief and practice while finding new significance in and new relationships between many others.[7]

Rigid standards from the nineteenth century institutionalization of science, for instance, underlie Kuhn's remarks on the role of "puzzle solving" or undergoing a rigorous training before one can effectively produce what are counted as "discoveries." For Kuhn, "sciences are not born de novo."[8] Thus, the role of professionalization in this sense is to underscore how the alternative approaches to major risk issues are developed so that they may be systemically explained, as through journal articles and books.

As provided here, this view does not cover the complex issues in the sociology and psychology of professionals and experts, including the propensity in some disciplines for significant errors. Instead, what is underscored is that alternative approaches that are acceptable must be relatively coherent, with a systemic development, and not merely opinions, or possibilities.[9]

---

[6] See, for instance, pp. 225, 234 in Kuhn, Thomas, 1977, op.cit.

[7] See pp. 226–227 in Kuhn, Thomas, 1977, Ibid.

[8] See p. 234 in Kuhn, Thomas, Ibid.

[9] On pp. 112–113, in Flyvbjerg, Bent, Nils Bruzelius, and Werner Rothegatter, 2003, *Megaprojects and Risk: An Anatomy of Ambition*, Cambridge, UK: Cambridge University Press, Funtowicz and Ravetz are used to define "peers" either as scientists and experts who are colleagues working within the "paradigm of the official expertise" or those who are enriched "at the very least" by the contribution of other scientists and experts and who are "technically competent but representing interests outside the social paradigm of the official expertise." In Flyvbjerg, Bent, 2001, *Making Social Science Matter: Why social inquiry fails and how it can succeed again*, Cambridge, UK: Cambridge University Press, Hubert Dreyfus is used on pp. 16–18 to define (level 4) "proficient performer: beyond analytical rationality" and "expert." Experts proceed intuitively, synchronically, and holistically to achieve a high level in situations in order to make decisions. Of course, there are libraries of works that show that experts can make mistakes, and these include mistakes on large-scale political judgments as shown in Tetlock, P. E., 2005, *Expert political judgment*, Princeton, N. J.: Princeton University Press. The definitions that Flyvbjerg, Dreyfus, and colleagues used do not require that experts be infallible. One finds the 10,000 h rule used for outlier performance in

Question 2: how does education and training provide a world view that assists in "bridging the gap" between finite samples and infinite populations—even though the bridges are corrigible? This second question comes from a very different perspective, if standard statistical approaches do not work well, how can one rely on any of them made by professional investigations? How can these individual investigations prove to work if the issue is how to bridge the gap between finite samples and infinite populations? Even 40 or 50 diverse investigations do not seem to be able to bridge the gap between finite samples and infinite populations.

One response to the second issue begins with the assumption—based on past evaluations of catastrophe risk evaluations—that many of the risk evaluations are going to be "Gaussian" in the broad sense of having catastrophe indexes above 2 and sometimes barely above 2.[10] Thus, the many tools that are common in statistics can be employed with some restraint but with the anticipation that the trajectory of outcomes will be sufficiently stable. This applies to results of individual evaluations.

A second response, perhaps more fruitful, resides in the view that a considerable portion of the inherited science that Kuhn regards as requiring "puzzle solving" consists of idealizations and heuristics. These idealizations and heuristics are essential to the building up of systemic knowledge; knowledge by its nature cannot be absorbed all at once. In referencing the "knowledge" of a physical law, Wittgenstein writes:

> "If the parts [e.g, of a crosshead] were quite rigid this is how they would move"; is that a hypothesis? It seems not. For when we say: "Kinematics describes the movements of the mechanism on the assumption that its parts are completely rigid", on the one hand we are admitting that this assumption never square with reality, and on the other hand it is not supposed to be in any way doubtful that completely rigid parts would move in this way. But whence this certainty? The question here is not really one of certainty, but of something stipulated by us.[11]

The "puzzle solving" of which Kuhn speaks can consist of finding solutions to old solutions, "divergences of ages past," that once took considerable more time and effort to find. For Kuhn, "in the mature sciences, most things generally do go right."[12]

Gladwell, Malcolm, 2008, *Outliers: The Story of Success*, New York: Back Bay Books, Little Brown and Company. Of course, there are many who have spent 10,000 h on some tasks and still not achieved a high level of competence.

[10] For this view for earthquake portfolio evaluations, see J. Lemaire and C. Tillman. 1993, "Models for Earthquake Insurance and Reinsurance Evaluations," Proceedings of the Second International Symposium on Uncertainty Modeling and Analysis, Los Alamitos, California: IEEE Computer Society Press, April.

[11] From p. 37e in Wittgenstein, Ludwig, 1967, *Remarks on the Foundations of Mathematics*, Cambridge, MA: the M. I. T. Press, first published in 1956, edited by G. H. von Wright, R. Rhees, G. E. M. Anscombe, and translated by G. E. M. Anscombe.

[12] See p. 222 in Kuhn, Thomas, 1977, op. cit.

The systemic view of knowledge and convergent knowledge underlying divergent findings leads to different views of "confirmation" and "falsification" from those in traditional statistics, on the one hand, and Popper's writings, on the other hand. Confirmation through data can support preexisting knowledge, or these data can upset preexisting knowledge. "Knowledge" in this sense is not something eternal but instead undergoes modification through research, experience, and application.

Hence, part of the answer here to the second question is that systemic knowledge does not "follow the rules" of incremental knowledge. This knowledge may contain idealizations and heuristics whose origins may not be in serious question as data are developed on a specific topic. This learned systematic knowledge may be very but not wholly coherent: errors may be embedded in what is currently known. Idealizations and heuristics may include the mathematical notion of "infinity" as used in probability and statistics. In particular and relative to finite samples, "infinity" may provide a picture of the trajectory or trajectories that a trend is leading. The catastrophe index in Chap. 5 provides trajectories based on finite data samples. Parametric models such as normal, lognormal, exponential, Poisson, and Pareto distributions can suggest trajectories of finite samples as they are increased.

In many instances, one does not use probability and statistics in order to find an ultimate trajectory or trend that goes to infinity. For instance, ballplayers wear out as do tires and machines. For individuals and tires, performance statistics may be very useful for a limited time horizon, but they are not going to continue forever.

Question 3: **how does one deal with such "deterministic" subjects as chaos theory when unknown initial conditions provide room for developing probabilistic models**? Where do subjects like classical mechanics, but also chaos theory, fit into the views presented so far? For instance, why has chaos theory not yet benefited from the understanding by Werner Heisenberg that: "When one wishes to calculate 'the future' from 'the present' one can only get statistical results, since one can never discover every detail of the present."[13]

The third question gives rise to some of a very large number of inquiries that can be undertaken. These include, for instance:

- Chaos theory has shown that "systems" can yield diverse and sometimes divergent dynamic trajectories. However, chaos theory has been primarily deterministic.[15] (chaos theory from Wikipedia) To what extent can one develop stochastic models of chaos theory? Can "initial conditions" be modeled as having endogenous uncertainty distributions?
- To what extent is there a currently inherent randomness in various highly numerical subjects such as geophysics, climatology, structural engineering, and cosmogony?

---

[13] Quoted on p. 333 in Isaacson, Walter, 2007, *Einstein: His Life and Universe*, New York: Simon & Schuster.

- What is the grain of truth in Nassim Taleb's view that distributions in the "Black Swan zone" are "incomputable"?[14] Can one accept the slightly contradictory or at least limiting view that extreme value distributions can be very useful, as in electrical engineering, yet ignore the "wobble" or merely regard the wobble as a minor issue relative to the value of using such extreme value distributions?
- Not discussed thoroughly are issues of nonlinearity. The discussion of catastrophe modeling of systems brings in nonlinearity for a great many systems. As Chap. 5 shows, one can develop a comparison between power laws and the catastrophe indexes. This suggests again that some systems have risks that are at best only very poorly computable. As Chap. 5 again shows, this may mean in the more extreme cases that the mode of losses for these systems is very well behaved but that the extremes rule out estimating statistical variances and even possibly arithmetic means.[15] Chapter 7 emphasizes the nonlinearity in addressing major healthcare issues. Nonlinearity is used in Chap. 7 to show how the long time frames in the developments of eventually successful working theories provide time for competing theories to thrive *even if* in rare cases the successful outcome remains so for an enormously long time.
- Also not explored enough is the so-called fallacy of affirming the consequence. This is indeed a fallacy when one has a single consequence. Formally speaking, this reads "C is true. If A were true, C would be true. Hence, there is reason to believe that A is true."[16] This formalization presupposes that one should stamp "T" or "F" on individual statements in their first instantiation, as though this was the last of a long narrative of movements backward and forward. As Chap. 7 has shown, with critical reasoning, a given theory may have a great many consequences that help to confirm the theory. The theory is not necessarily abandoned in toto because it has one or even many troublesome consequences. As shown in many examples, language may be changed with new additions as the discovery is clarified. The original "discovery" may undergo many changes as the initial discovery comes to terms with many issues. As with the continental drift theory (or a great many others such as the discussion of Kepler's discovery), the theory may be required to morph into a more comprehensive theory that overcomes many of the initial objections and provides clarification otherwise not available by other extant theories.

[14] See pp. 138, 288 in Taleb, Nassim Nicholas, 2012, *Antifragile: Things that Gain from Disorder*, New York: Random House.

[15] For describing chaos as entering when the number of variables in a nonlinear model exceeds 3 and also when the topic has yet to be explored enough, as with immune systems and ecosystems, see p. 11, Strogatz, Steven H., 1994, *Nonlinear Dynamics and Chaos: With Applications to Physics, Biology, Chemistry, and Engineering*, Cambridge, MA: Perseus Books Publishing, LLC.

[16] These simplifications are similar to the formalizations in Peirce, Charles, 1901, 1903, in "Abduction and Induction," pp. 150–156; in Charles Peirce, 1955, ibid., p. 151; and Wikipedia, "Abductive reasoning," accessed 12/14/2013. Note that Peirce's discussion of Kepler's discoveries, which are not easily encapsulated in two or three lines, compares with the discussions of discoveries in healthcare discussed in Chap. 7.

Theories entitled to be considered comprehensive views are systemic in the sense that a single negative consequence (or datum that oppose the theory) does not eliminate the theory. Instead, negative consequences may provide opportunities for further modification of a comprehensive theory, as has happened for well over a century with Darwin's theories.[17]

Question 4: **from a program management perspective, or in general, how does one encourage competition among investigators, how does one assemble alternative outcomes in ways that are useful to decision-makers, and what happens after these outcomes are assembled**? In short, how does one comprehend the phenomenology of this competition among competitors (e.g., are alternatives exhausted in some cases, is institutional memory lost, is there in some cases convergence resulting from later investigations, are resources exhausted, do new competitors arise and how)?

Question 5: **how does one test the value of tools described in Chap.8with respect to their value in making decisions regarding mega-risks**? How does one provide these when the multiple outcomes are "deterministic"? Is the principle of least regret adequate for all such applications? And, how does one provide a quantitative account of decision analysis for ensemble statistical outcomes? Previously there have been stochastic accounts that have been provided using mean values, statistical variances, the entire distribution of gains and losses (stochastic dominance), the almost stochastic dominance, fat-tail reduction models, and even the use of multiple decision criteria.

9.4 Conclusions

Robust simulation provides a largely coherent response to the question as to how one can account for uncertainties in complex risk evaluations. In practice, different statistical results of professional investigations have been presented in many important disciplines. Some of these are found in the social sciences and some are in highly quantitative physical sciences, especially as computational speed and capacity permit alternative search and other approaches to major issues. The vast number of unknowns or matters poorly known gives rise in socially important risk issues to a process that enables diverse investigative teams to engage in a process of deconstructing previous views, construction alternative views, experimenting further, deconstructing the newer views, and reconstructing even more preferred views. Even after long periods of inquiry and large data samples, these resulting views may still differ among different investigative teams.

[17] These simplifications are similar to the formalizations in Peirce, Charles, 1901, 1903, in "Abduction and Induction," pp. 150–156; in Charles Peirce, 1955, ibid., p. 151; and Wikipedia, "Abductive reasoning," accessed 12/14/2013. Note that Peirce's discussion of Kepler's discoveries, which are not easily encapsulated in two or three lines, compares with the discussions of discoveries in healthcare discussed in Chap. 7.

Robust simulation proves an answer as to how to calculate the uncertainties in these risk-related issues. Robust simulation also requires that the alternative approaches used be credible, coherent, or professional. As a result of nonlinearity in developing more definitive views, these approaches imply taking into account years of experience in the topics in question. Active minds are required to develop approaches that can be used. Robust simulation uses rather than totally displaces alternative statistical approaches, even those that have traditionally required only unique statistical solutions.

These essays provide hints as to how there may be a bridge between experience and the infinite populations typically assumed in probability and statistics. The systematization of knowledge, including idealizations and heuristics, provides a basic background for how disciplined experience is not confined to finite samples. The use of a community of investigators also emphasizes the role of active research and application in the development of an ensemble of outcomes for socially important risk issues. This community is "involved in an iterative process, questions going back and forth, not a single answer, but interaction and useful representation of uncertainties."[18]

References

Bernstein, P. L. (1996). *Against the Gods: The remarkable story of risk*. New York: Wiley.

Flyvbjerg, B. (2001). *Making social science matter: Why social inquiry fails and how it can succeed again*. Cambridge: Cambridge University Press.

Flyvbjerg, B., Bruzelius, N., & Rothegatter, W. (2003). *Megaprojects and risk: An anatomy of ambition*. Cambridge: Cambridge University Press.

Gladwell, M. (2008). *Outliers: The story of success*. New York: Back Bay Books, Little Brown and Company.

Isaacson, W. (2007). *Einstein: His life and universe*. New York: Simon & Schuster.

Kuhn, T. (1977). *The essential tension: Selected studies in scientific tradition and change*. Chicago: University of Chicago Press.

Lemaire, J. & Tillman, C. (1993, April). Models for earthquake insurance and reinsurance evaluations. In: *Proceedings of the second international symposium on uncertainty modeling and analysis*. Los Alamitos: IEEE Computer Society Press.

Nolan, J. P. (2009). *Stable distributions: Models for heavy tailed data*. Accessed on the internet 27 Feb 2013.

Peirce, C. (1901, 1903). In abduction and induction. In Peirce Edition Project (ed.), *The essential Peirce, volume 2: Selected philosophical writings*, (pp. 1893–1913). Bloomington, Indiana University Press.

Strogatz, S. H. (1994). *Nonlinear dynamics and chaos: With applications to physics, biology, chemistry, and engineering*. Cambridge, MA: Perseus Books Publishing, LLC.

Taleb, N. N. (2012). *Antifragile: Things that gain from disorder*. New York: Random House.

Taylor, C., Murnane, R., Graf, W., & Lee, Y. (2013). Epistemic uncertainty, rival models, and closure. *Natural Hazards Review, 14*(1), 42–51.

---

[18] Written communication, Melissa Taylor Dresler, 7/22/13.

Tetlock, P. E. (2005). *Expert political judgment*. Princeton: Princeton University Press.
Toulmin, S. (1992). *Cosmopolis: The hidden agenda of modernity*. Chicago: University of Chicago Press.
Wittgenstein, L. (1967). *Remarks on the foundations of mathematics*, Cambridge, MA: the M. I. T. Press, first published in 1956, edited by G. H. von Wright, R. Rhees, G. E. M. Anscombe, and translated by G. E. M. Anscombe.

Essay: Learning from Tradition (Original Preface)

The Contrasting Trajectory in This Work: Learning from Traditional Sources of the Single-Solution Approaches, Leading to Competitive, Multiple Interpretive Solutions

> Ideas in modern Russian are machine-cut blocks coming in solid colors; the nuance is outlawed, the interval walled up, the curve grossly stepped.[1]

> [goal]: to transform social science from what is fast becoming a sterile academic activity which is undertaken mostly for its own sake and in increasing isolation from a society on which it has little effect and from which it gets little appreciation.[2]

Stressing the Phenomena of Mega-Risks and Importance of Multiple Interpretations

This work is about mega-risks. These include natural and technological disasters, large-scale construction projects, missile and asteroid risks, political risks including wars and other violent conflicts, climate change risks, and epidemics. Frankly speaking, no one knows much about these mega-risks, yet many of us have spent many hours on one or more of these phenomena. This work chiefly covers quantitative evaluations and by a presumptive extension considers these multiple interpretations when there are more than one valid evaluations at a given time.

Another presumption throughout this document is that multiple interpretations are most desirable whenever imagination or novelty is needed. Of course imagination is required on a daily basis in order to consider alternative courses of action as

---

[1] From *Pale Fire*, 1962, by Vladimir Nabokov, New York: Perigee Books. p. 243.

[2] From p. 166 in Flyvbjerg, Bent, 2001, *Making Social Science Matter: Why social inquiry fails and how it can succeed again*, Cambridge, UK: Cambridge University Press.

C.E. Taylor, *Robust Simulation for Mega-Risks*, DOI 10.1007/978-3-319-19413-4

desired or needed. Contrary to the second quotation above, one might maintain that a major function of the humanities and social sciences is to provide the ability to look at matters imaginatively, in different ways. This document, though, stresses how in many areas—starting with "hard" sciences—the predominant way of thinking has presumed that there is but one interpretation, one evaluation, and one solution. If this thought were confined to "hard" sciences, then the discovery or clarification that multiple interpretations are available even in "hard" sciences would go some way toward bridging the gap in what has been deemed the issue of "two cultures," if indeed the second quotation above did not have more than a grain of truth. So, for this preface, one might pursue this thought for a few pages. This document is also premised on the notion that, although modeling and assessment efforts have uncertainties, they provide value in risk decision-making.

This preface deals with alternative trajectories that have been taken on related issues and how these contrast to the essays in this book.

Diverse past Trajectories on These Issues: Those Stressing Lack of Nuance in the Social Sciences, Those Learned in the Humanities Who Exalt the Social Benefits of the "Harder" Disciplines with Their Rigid Methods

Seeking for similarities in past discussions, I found only very different trajectories to related issues are found in John Stuart Mill's essay on Bentham and Coleridge, C.P. Snow's essay on two cultures, and Henry Adams' discussion of changes in US culture from one emphasizing the humanities to one stressing industrialization.

In 1838 and 1840, John Stuart Mill provides a classical discussion of the distinction between two thinkers: Jeremy Bentham and Samuel Taylor Coleridge. Mill regarded each thinker as leaders in questioning the status quo and as having enormous influence on the British culture of the day. Summarily, Mill regards Bentham as providing a "scientific" or empiricist method of how to approach business, legal, and political issues. This empiricist and extreme utilitarian view regarded single outcomes as the result of a "method of exhaustion" in which one viewed all alternatives and ruled out all but one, which was the truth. For Mill, Bentham's scientific approach was valid, but his genuine contributions lay in business and law. Bentham lacked the imagination of being able to understand what others thought or taking their opinions seriously.

In contrast, Coleridge emphasized the importance of mining what others thought, in assuming that what underlay the opinions of others contained some significant merit. Thus, Coleridge was concerned with exploring the meanings of what others' past and present had derived and taking them seriously.[3]

[3] Pertinent pages include pp. 41, 44, 48, 56, 62, and 75 in 1838, "Bentham," and pp. 99, 100, and 143 in 1840, "Coleridge" in *Mill on Bentham and Coleridge*. One very faint clue that the social

Mill's discussion stresses how someone in the humanities, namely, Coleridge, can be principally concerned with issues in interpretations. At the same time, like the two quotations that lead this preface, Mill's discussion thus confirms that there is a view of the social sciences that lacks nuance—that stresses that there is but one solution, one result. Thus, some or much social science work may lend itself to being more "quantitative." From this standpoint, the "two cultures" is more complex. For a variety of reasons, including the resources provided for the "hard" culture, this "hard" culture may on occasion be far more imaginative in exploring various solutions. Writings on "two cultures" typically do not consider this alternative.

C.P. Snow's short work on *The Two Cultures* does not bridge this gap. Instead, this work provides an apology for all the advantages that industrialization and science have provided so that those in the humanities should worship at its altar. Snow applauds how science has removed "unnecessary suffering from a billion individual human lives":

> The scientific revolution is the only method by which most people can gain the primal things (years of life, freedom from hunger, survival for children).[4]

For Snow, the scientific method and its industrial application have had all these obvious advantages that the Luddite tendencies lead those in the humanities to fail to understand. The common culture is for Snow rightly dominated by the scientific-technical culture that has provided these advantages.

Based on my first reading of *The Education of Henry Adams* by Henry Adams, I thought that this work might provide a clue as to how the humanist culture in which Henry Adams had been raised largely before 1850 had intertwined with the industrialist culture afterward and how he had found a way to reconcile the two cultures. In effect, I thought that Adams would provide a better view as to how the two cultures, the heavily quantitative and the more qualitative, weave together in at least a slightly stable harmony.

In his delightful work, Adams though tends to suffer from the same trajectory as does Snow: the person educated in the classics and broad humanities now lauds principally the quantitative approaches that underlie the industrial society. This becomes evident in the view of history that Adams sketches, a view derived not from the humanities nor from the life sciences but instead from nineteenth-century physics. Adams proclaims a kinetic theory of history, a theory determined by an acceleration resulting from a new-found energy and economy of forces. In 1840, The trusts and corporations were created that stood for the larger part of the new power. These trusts and corporations had a "vigorous and unscrupulous energy." They troubled "all the old conventions and values, as the screws of ocean streamers

sciences may not be the more enlightened as regards multiple interpretations lies in the almost total absence of a discussion of "skewness" in the excellent work by William L. Hays, 1973, *Statistics for the Social Sciences*, New York: Holt Rinehart and Winston, Inc. As the document that follows points out, skewness is extremely important in evaluating mega-risks.

[4] From pp. 78 and 80 in Snow, C. P., 1963, *The Two Cultures and A Second Look*, Cambridge: Cambridge University Press. See also pp. 3, 22, 60, and 70.

must trouble a school of herring. They tore society to pieces and trampled it under foot." The new world consisted of people who rejected theories of history, philosophy, and theology but instead regarded the world as being full of new energies.[5]

By 1900 not long before Adams died, this energy consisted of coal power, electric power, and radiating energy. This energy, along with new forces as yet undiscovered, might enable people to forecast changes with ever increasing accelerations and velocities, which, if plotted, might yield predictions to plot the past and future orbit of the human race as accurately as that of the November meteoroids. The old qualitative methods of dispute and discussion had become idle in the face of these new energies and their attendant forces.[6]

The Contrasting Trajectory in This Work: First, Learning from Traditional Sources of the Single-Solution Approaches to Mega-Risks

The trajectory of the following essays goes in the opposite direction from the works of Mill, Adams, and Snow. These essays do not start from eminently readable discussions of society, education, politics, the humanities, and diverse viewpoints therein. These essays do not end with an affirmation that the quantitative disciples are fait accompli and should provide the forces that leave disputes and discussion worthless. These essays do not deny that quantitative approaches have in many cases dominated discussions, innovations, inventions, and everyday life.

Instead, after defining the issues at hand and summarizing chapters to come, these essays begin with highly quantitative approaches of statistical and probability theories that now make a considerable difference in everyday life. These essays begin with discussions of a "deductivist" approach, proceed to a "frequentist" approach, and continue with a "Bayesian" approach. These three approaches have traditionally been the main approaches to issues in probability and statistics, and the latter two approaches still dominate today's discussions. These discussions are designed to address questions about accounting for uncertainties in mega-risk evaluations. In spite of all the valuable insights that come from these three approaches, they address uncertainty issues in very limited ways. Their limitations are developed in these chapters as well as in the next chapter that provides a simple approach to the initial assessment of extreme value distributions.

To complete the discussion of how traditional theories fare in addressing uncertainty issues, a discussion of the mathematization of probability and statistics follows. Once again, the discussions first consider quantitative issues and of how they

[5] From pp. 239 and 500 in Adams, Henry, 1918, *The Education of Henry Adams*, Boston, Houghton Mifflin Company.

[6] From pp. 496 and 501 Adams, Henry, op. cit.

have been treated and how they could be treated. The mathematization of statistics purports to bridge the gap between finite samples and the infinite populations used in probability and statistics. The vast improvements in information technology (IT) have led to and can continue to lead to many significant results in probability and statistics because one is enabled to visualize how huge numbers of simulations might yield sound results. Yet, very briefly speaking, when one sees in one's mind's eye what an infinity of samples might be, one finds the passage from Shakespeare's *Troilus and Cressida* apropos:

> The will is infinite
> and the execution confined,
> the desire is boundless
> and the act a slave to limit.[7]

These chapters are not dead ends because each chapter contains explicitly and implicitly enormous advances in the infrastructure of probability and statistics and assists in addressing major questions on stochastic uncertainties in evaluations of mega-risks. But the trajectory still has a distance to traverse.

The Contrasting Trajectory in This Work: Interpretive Processes Dealing with Mega-Risks in Competitive Highly Disciplined Settings

The same is actually true with, for instance, reading J.S. Mill, C.P. Snow, and Henry Adams, respectively. One doesn't find out how they are coping with the issue of two cultures or the industrialization (and now the post-industrialization) of society and culture, given previous educational emphases away from quantitative disciplines currently favored. My interpretation of J. S. Mill, C.P. Snow, and Henry Adams, respectively, may turn out on further examination to be way off base. I use the term "interpretation" in the title because it is a term used in many "softer" disciplines and is not thought of as being as meaningful in the more strict quantitative disciplines. "Interpretation" requires the "imagination," defined by Mill as conceiving the absent as if it were present and the imaginary as if it were real, and, by implications, makes it possible to interpret another mind, way of thinking, or approach to dealing with a problem.[8] What is treated as imaginary may turn out to be real or a valuable way to solve some major issues.

So, each new chapter is an outgrowth of findings from the previous chapters. The final three chapters produce an infrastructure for the treatment of uncertainties that is much richer than that achieved by thinking of uncertainties as being dealt with by large samples. Instead, the final chapters rely more significantly on the discipline developed through education and training that enables one to make sound or intel-

[7] From p. 38 in Mandelbrot, Benoit B., 1983, *The Fractal Geometry of Nature*, New York: W. H. Freeman and Company, originally 1977.

ligent decisions that are stochastic in nature. This discipline can be combined with a critical account of existing quantitative approaches that may be useful in making decisions about mega-risks. These quantitative decision procedures do not rule out numerous qualitative considerations nor the issue of interpretation.

This trajectory thus proceeds in the opposite direction from that of J. S. Mill, Henry Adams, and C.P. Snow but finds similarities in the writings of many others. The trajectory in these essays moves more toward a "soft" approach to uncertainty in these vast simulations as "harder" approaches one after another fall by the way-side. We "interpret" outputs from evaluating shocks to systems, and this interpretation, fallible as it is, should require years of prior discipline and appeal to those in various disciplines. Some problems can be addressed in an instant; to address other problems may require decades or longer. In the words of William James in 1882:

> A chemist who conjectures that a certain wall-paper contains arsenic; and has faith enough to lead him to take the trouble to put some of it into a hydrogen bottle, finds out by the results of his action whether he was right or wrong. But theories like that of Darwin, or that of the kinetic constitution of matter, may exhaust the labors of generations in their corroboration, each tester of their truth proceeding in the simple way—that he acts as if it were true and expects the results to disappoint him if his assumption is false.[8]

The trajectory in Flyvbjerg quoted at the start of this preface is first to indicate how science and scientific activities follow something more like a Cartesian or Newtonian viewpoint or like Thomas Kuhn's "normal" science and Aristotle's view of science. Flyvbjerg contrasts a context-free science emphasizing universals and a cumulative view of progress to his view of the social sciences as a modification of Aristotle's theory of *phronesis* or practical reasoning that emphasizes context-dependent concrete activities in valuation settings. "Interpretation" becomes critical in Flyvbjerg's view of social science. Without going further into his view of social science, there is to some degree more of a merger here of his view and the view in these essays. In particular, the so-called "hard" sciences are when applied to challenging issues far less "Newtonian" than classical physics defined as a "puzzle-solving" activity in which there is one solution to each puzzle. The quantitative areas considered in these essays naturally give way to numerous quantitative approaches to decision-making and to that extent have bearing on valuation issues. These quantitative approaches to decision-making assist but do not in these essays purport to deal wholly with all the concrete activities involved in decision-making.

The view here furthers the work of Imre Lakatos, who maintains that "normal science" is nothing but a research program that has achieved monopoly"[9] and "[*t*]*he history of science has been and should be a history of competing research programmes*."[10] The systemic and contextual features of "puzzle solving" are for-

[8] From p. 325, William James, 1882, "the Sentiment of Rationality," in *The Writings of William James*: *A Comprehensive Edition*.

[9] From p. 69 in Lakatos, I., 1978, *The methodology of scientific research programmes*, London: Cambridge University Press.

[10] From p. 69 in Lakatos, I., 1978, *The methodology of scientific research programmes*, London: Cambridge University Press.

gotten when one ignores how one "law" may interact with others and be part of a puzzle: the ball rolls uphill as a result of the westerly wind. So, when a new force is discovered, there is a systemic modification of the previous view.

The trajectory in these essays is indeed more like the trajectory in the work of the extremely divergent thinking Benoit Mandelbrot, whose views have not been fully absorbed in these essays. Although one of his first discoveries was the price changes in economics did not follow the Gaussian distribution as had been theorized since the turn of the century, this theory continued to be used in spite of the failure of Long-Term Capital Management and until perhaps the great recession of 2008, so, he did much work in "harder" sciences and engineering such as astrophysics on the distribution of galaxies, hydrology on different impacts of dams on rivers, and electrical engineering to evaluate turbulence in flows.[11]

References

Adams, H. (1918). *The education of Henry Adams*. Boston: Houghton Mifflin Company.

Coleridge. (1840). In *Mill on Bentham and Coleridge (1950) with an Introduction by F.R. Leavis*. Chatto and Windus.

Flyvbjerg, B. (2001). *Making social science matter: Why social inquiry fails and how it can succeed again*. Cambridge: Cambridge University Press.

Hays, W. L. (1973). *Statistics for the social sciences*. New York: Holt Rinehart and Winston.

James, W. (1882). The sentiment of rationality. In *The writings of William James: A comprehensive edition*. Phoenix Book.

Lakatos, I. (1978). *The methodology of scientific research programmes*. London: Cambridge University Press.

Lowenstein, R. (2000). *When genius failed: The rise and fall of long-term capital management*. New York: Random House.

Mandelbrot, B. B. (1983). *The fractal geometry of nature*. New York: W. H. Freeman and Company. originally 1977.

Mandelbrot, B. B. (2012). *The fractalist: Memoir of a scientific Maverick*. New York: Pantheon Books.

Mill, J. S. Bentham. In *Mill on Bentham and Coleridge (1950) with an Introduction by F.R. Leavis*. Chatto and Windus.

Snow, C. P. (1963). *The two cultures and a second look*. Cambridge: Cambridge University Press.

Vladimir, N. (1962). *Fire, pale*. New York: Perigee Books.

[11] Mandelbrot, Benoit B., 2012, *The Fractalist: Memoir of a Scientific Maverick*, New York: Pantheon Books. On failures on Long-Term Capital Management and the theories that were severely damaged, see Lowenstein, R., 2000, *When Genius Failed: The Rise and Fall of Long-Term Capital Management*, New York: Random House.

Index

C.E. Taylor, *Robust Simulation for Mega-Risks*, DOI 10.1007/978-3-319-19413-4

MIX
Papier | Fördert gute Waldnutzung
FSC® C083411

Kolibri 360 GmbH
Ferdinand-Jühlke-Straße 7
99095 Erfurt, Deutschland
produktsicherheit@kolibri360.de